Construction Site Management and Labor Productivity Improvement

Other Titles of Interest

Productivity Improvement for Construction and Engineering: Implementing Programs that Save Money and Time, **by J. K. Yates, Ph.D.** (ASCE Press, 2014). Focuses on investigation and analysis techniques that can be used by engineering and construction firms to support the implementation of productivity improvement programs.

Construction Contract Claims, Changes, and Dispute Resolution, Third Edition, **by Paul Levin.** (ASCE Press, 2016). Paul Levin and a team of construction experts and attorneys guide contractors, engineers, owners, and construction managers through the complex process of construction contracting, focusing on claims and change orders in construction projects.

Interpreting Construction Contracts: Fundamental Principles for Contractors, Project Managers, and Contract Administrators, **by H. Randolph Thomas, Ph.D., P.E., and Ralph D. Ellis Jr., Ph.D., P.E.** (ASCE Press, 2007). Discusses the most troublesome contract clauses and presents rules to construe them so as to avoid disputes that must be resolved in court.

Managing Gigaprojects: Advice from Those Who've Been There, Done That, **edited by Patricia D. Galloway, Ph.D., P.E.; Kris R. Nielsen, Ph.D., J.D.; and Jack L. Dignum.** (ASCE Press, 2012). A stellar group of financial, legal, and construction professionals share lessons learned and best practices developed from working on the world's biggest infrastructure construction projects.

Project Administration for Design-Build Contracts: A Primer for Owners, Engineers, and Contractors, **by James E. Koch, Ph.D., P.E.; Douglas D. Gransberg, Ph.D., P.E.; and Keith R. Molenaar, Ph.D.** (ASCE Press, 2010). Explains the basics of administering a design-build project after the contract has been awarded.

Public-Private Partnerships: Case Studies on Infrastructure Development, **by Sidney M. Levy.** (ASCE Press, 2011). Demystifies public-private partnerships as an innovative solution to the challenges of designing, financing, building, and operating major infrastructure projects.

Construction Site Management and Labor Productivity Improvement

How to Improve the Bottom Line and Shorten Project Schedules

H. Randolph Thomas, Ph.D., P.E.
Ralph D. Ellis Jr., Ph.D., P.E.

ASCE
PRESS

Library of Congress Cataloging-in-Publication Data

Names: Thomas, H. Randolph, 1945- author. | Ellis, Ralph D. (Civil engineer), author.
Title: Construction site management and labor productivity improvement : how to improve the bottom line and shorten project schedules / H. Randolph Thomas, Ph.D., P.E. Ralph D. Ellis, Jr., Ph.D., P.E.
Description: Reston, Virginia : American Society of Civil Engineers, 2017. | Includes bibliographical references and index.
Identifiers: LCCN 2017010873| ISBN 9780784414651 (print :alk. paper) | ISBN 9780784480328 (PDF) | ISBN 9780784480649 (ePUB)
Subjects: LCSH: Construction projects–Management. | Project management. | Building–Superintendence.
Classification: LCC TH438 .T485 2017 | DDC 624.068/5–dc23 LC record available at https://lccn.loc.gov/2017010873

Published by American Society of Civil Engineers
1801 Alexander Bell Drive
Reston, Virginia 20191-4382
www.asce.org/bookstore | ascelibrary.org

Errata: Errata, if any, can be found at https://doi.org/10.1061/9780784414651.

ISBN 978-0-7844-1465-1 (print)
ISBN 978-0-7844-8032-8 (PDF)
ISBN 978-0-7844-8064-9 (ePUB)
Manufactured in the United States of America.

24 23 22 21 20 19 18 17 1 2 3 4 5

Contents

Preface

This text is based on the collective 70-plus years of experience the authors have in monitoring and managing construction projects. During this time, the authors have monitored more than 200 active construction projects. On most of these projects, the labor productivity was measured, loss of productivity was noted, and the deficient events leading to productivity degradation were documented.

The authors have observed that these events can be organized in a relatively few broad categories, such as material management and sequencing. Furthermore, the authors have also observed that whenever these deficient events (conditions) are present, the consequences are always the same: inefficient productivity and extended schedules. It follows that the management practices to improve productivity and shorten schedules should strive to ensure that deficient events are not present on one's project. The principles cited in this book focus on managing site operations to improve construction labor productivity and shorten schedules. The management practices to avoid deficient events are simple and are based on common sense, not theory. The principles are things managers can readily and easily implement to manage site operations. This book is written primarily for practicing engineers and contractors. The management actions are labeled "fundamental principles." An example illustrating a fundamental principle is to "minimize the number of times a crane has to be moved." The fundamental principles can be easily organized into checklists.

Fundamental principles are different from best practices because with fundamental principles there is documented evidence that the application of the principle will lead to positive results. Best practices are simply that. They are things contractors do, and there may be no evidence to show that their application will lead to improvements.

This text is divided into four parts. The first part states the commonly reoccurring causes of inefficiency and examines what has been said by researchers and practitioners about these causes. Part II pertains to planning. A comprehensive, but straightforward, planning procedure is given, and a simple heuristic procedure is illustrated for developing an efficient site layout plan. Part III details fundamental principles for weather mitigation, material management, workforce management, activity and trade sequencing, avoiding congestion, and subcontractor management.

In Part IV, Special Topics, schedule acceleration and environmental compliance are discussed. The last chapter in Part IV explains how the fundamental principles can be used to evaluate a project.

In total, there are more than 95 fundamental principles. Each is explained in detail so that it can be correctly applied, the reason for the principle is known, and the expected outcome is clear. The principles are fully illustrated with case studies and photographs showing the deficient events the principle is intended to address. If these principles are judiciously applied, the authors believe that labor productivity can be improved by 15% and schedules can easily be improved by 20%.

PART I

Introduction

CHAPTER 1

Introduction

This text emphasizes a trend that, for many construction projects, time is crucial. In this scenario, the performance of others is an important concern of general contractors, who cannot afford to take a hands-off approach to subcontractors. A team approach is sought. The goal of this book is to provide practical information for professionals working as construction project engineers or project managers about how to improve productivity and schedule performance.

The text is full of practical and easy-to-implement principles and is written primarily for practicing engineers. The book should not be placed on a shelf to gather dust; it needs to be used often. Fundamental principles of planning and management for contractors are given. The emphasis is on site construction management. The outcomes of the application of the principles cited herein will be better cost and schedule performance and higher profits.

There are many ways to erode profits on a construction project; poor labor performance is one of the more common ways. Achieving good productivity and schedule performance requires the application of good management practices. Management is responsible for (1) providing the resources, such as materials, and (2) creating an environment that is conducive to efficient production of output. This book is based on these two elements, providing resources and creating the right environment. But what are good management practices in these two areas? The most commonly observed and recurring deficiencies in these two areas are covered in this book. The topics include planning, site layout, mitigating the effects of adverse weather, material management, workforce management, activity sequencing, trade sequencing, avoiding congestion, subcontractor management, and managing schedule acceleration.

Fig. 1-1 models the process of converting inputs to outputs. The left side of the figure shows the resources needed to efficiently convert inputs to outputs. Entire chapters are devoted to material and labor resources. The top portion of Fig. 1-1 shows that the proper environmental conditions must be present for the conversion to be efficient. There can be other environmental factors besides the four factors shown. Complete chapters are devoted to weather mitigation, supervision, congestion, and sequencing. Unfortunately, if any one of these areas is deficient, the conversion of inputs to outputs will be impeded, and there will be inefficient use of

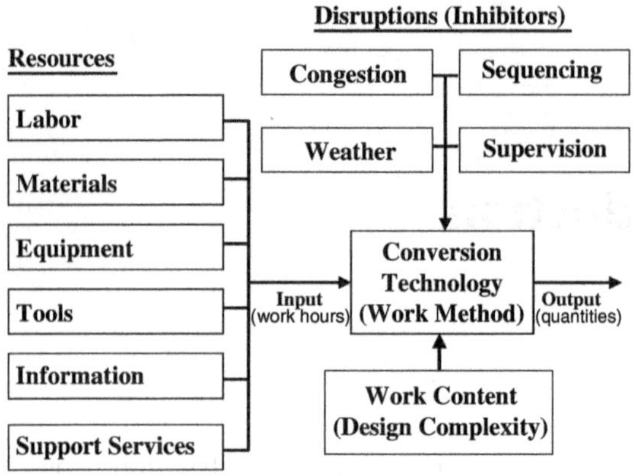

Fig. 1-1. The Factor-Resource Model

the labor resource or excessive idle time and reduced output. The fundamental principles in this book address the conditions that lead to inefficiencies and idle time. If the principles cited are applied, the U.S. workforce, which is among the best in the world, will produce stellar results.

Fig. 1-1 was developed from productivity monitoring activities on about 200 active construction projects. Losses of productivity have been calculated, and the causes of subpar productivity performance were noted for each project. Thus, Fig. 1-1 shows the most commonly recurring causes of productivity loss; it is cited frequently in the literature as the Factor-Resource Model. The Factor-Resource Model is the only productivity model that has been validated using actual field data.

1.1 Trends in Project Management

Construction project management has experienced two important trends in the past half-century that are partly the motivation for this text. First, time has become an increasingly important element of the contract exchange. Second, the role of project engineers has evolved to one of increasing significance. Both of these trends are discussed in the following.

Time as an Element of Performance

Some 50 years ago, cost was the primary consideration in project performance; time was a secondary consideration for many projects. Sweet and Schneier (2009), in his book on the legal aspects of construction, suggests that time and costs are treated differently. He says, "The law has *not* looked at time as part of the *basic* construction contract exchange—that is, money in exchange for the project." However, from a practical standpoint, that stance seems to have changed in recent years. Owners want timely completion more than ever. Certain types of projects are particularly time sensitive.

These include school projects, some apartment complexes or condominiums, sports stadiums, selected industrial and manufacturing projects, and wastewater facilities and other types of projects that may be under a court order. Court orders, loss of revenue, and the monetary cost seem to be three important factors driving this trend. This book emphasizes the efficient use of labor and timely completion. Contractors should strive to complete projects in as short a time as practical, thereby minimizing the time of exposure to certain risks.

Changing Role of the Project Engineer

A half-century ago, there were few project engineers, and many of the tasks described in this text were performed by construction superintendents. But over time, projects have become more complex, the monetary cost has become more significant, wage rates have risen, and project completion is a more significant aspect of project performance. Many of the functions formerly performed by superintendents are now assigned to project engineers. Additionally, in some locales, owners are insisting that the project manager be a licensed engineer or certified by a relevant professional organization. Universities have responded by graduating more than 1,000 project engineers annually. But many of these engineers do not know the basics of how to achieve good labor performance. Thus, the topics discussed in this book are pertinent to project engineers and project managers.

1.2 Labor Performance and Management Practices

The definition of *labor productivity* used in this text is the work hours per unit of work. Better productivity translates to better cost performance and better schedule performance. Good productivity and schedule performance on a construction site requires the application of good management practices. Sadly, many contractors do not demonstrate good practices. In a lifetime of monitoring active construction projects, the authors have observed few projects where good management practices were consistently applied. On many projects, good practices are applied in some respects, but there is another area of the project that is deficient. This deficient area is sufficient to lower job performance to average or mediocre levels. Project engineers and project managers must manage all aspects of the work.

Focus Is on Good Site Management

To improve labor productivity and competitiveness in the construction industry, there is a real need to assemble the body of knowledge of good site construction management practices in one place and add new practices as the need arises. This book is an effort to satisfy this goal. The authors believe that through the application of the fundamental principles detailed in the following chapters, labor productivity can easily be improved by 15% and project schedules can be shortened by 20%.

Methodology for Developing Fundamental Principles

The principles presented in this book were developed using a simplistic process. Over the past 30 years, the authors have monitored and measured productivity on more than 200 active construction projects. About 15% of these projects were international. The factors affecting labor productivity were documented, and the labor inefficiencies were calculated. It was observed that whenever certain things happened, the effect on labor productivity was always negative. A few factors had a positive effect. The authors are proposing practices intended to block the negative factors and promote the positive factors. Some of the practices were proposed by industry professionals, but most are just common sense. Other practices have been reviewed by industry professionals.

None of the practices are theoretical. All the practices are easy to understand and implement. The practices are almost sure to yield positive results.

1.3 Approach

It is recognized that the general subject of project management is quite broad, and it is impractical to address this broad topic in a single text. The authors approach this dilemma as follows.

This book does not try to address the entire subject of project management. Instead, the book is focused on site management only. This book is further limited because it focuses largely on labor usage, the most risky resource consumed on the site. Most writings on labor productivity present a rather complex view of labor productivity. This book presents a simplified view of labor productivity. Based on 30 or more years of extensive field research, the authors have concluded that eight topical areas that occur repeatedly adversely affect productivity, and only these areas are discussed in this book. Chapters 5–12 are devoted to these topical areas. The text presents inexpensive management actions that can be taken by project engineers and managers. Topics such as labor relations, improvement programs, and human factors are not discussed. This book is not written for beginners; it is assumed that readers have some rudimentary knowledge of what occurs on worksites. Therefore, elementary topics are not discussed. Instead, the text focuses on topics that are known, but may have been given little or no thought.

Most chapters begin with some limited background information on familiar topics. The text is thought-provoking in a way that likely has not been considered. The next part of the chapters presents fundamental principles or best practices. The principles are specific things a project engineer can do to improve labor performance, such as seal exposed areas each day, use a 4-10 work schedule (in which two teams work four 10-hour days) to permit a weather makeup day on straight time, or backfill around buildings as soon as practical. There are more than 95 principles contained in the text. Though many principles appear trite and elementary, research has shown that many of these principles are not practiced. The chapters conclude with case studies that illustrate the principles being discussed and the negative effect on labor productivity when various principles are ignored. The chapters are concisely written, and extensive figures and photographs are used in lieu of voluminous text.

1.4 Fundamental Principles and Best Practices

There is a difference between fundamental principles and best practices. For the projects monitored in detail, it was determined that there was a loss of efficiency attributed to a given practice. For example, on one project, it was determined that the loss of efficiency attributed to the practice of using the construction site as a staging area for structural steel erection resulted in a loss of 109 out of 768 work hours or a 16.5% loss of efficiency. Similar losses were calculated for other projects whenever the same practice was applied. It is concluded that whenever this particular practice is applied, labor is inefficiently used. Following discussions with contractors, it was decided that structural steel erection directly from the delivery truck is an efficient delivery method. Erection from the delivery truck is a fundamental principle because there is quantifiable evidence that erecting from the delivery truck is efficient and saves money by increasing labor efficiency.

Conversely, in regard to subcontractor management, engaging in the practice of using a subcontractor to assist in developing a prebid schedule is simply a best practice that is recommended. The authors have no documented evidence that this practice saves money. Best practices merely report what contractors presently do, good or otherwise.

Throughout the description of fundamental principles in this book, no distinction is made between the labeling of a practice as a fundamental principle or a best practice, but most practices are fundamental principles. The entire chapter on subcontractor management is composed of best practices, as is some of the chapter on planning, yet these chapters are titled fundamental principles.

1.5 "Not on My Project"

This text illustrates and describes undesirable situations on numerous projects. As an example, the reader is invited to review the photographs in Chapters 4–11. There is much to be learned from projects that do not go as hoped or expected. The puzzling question is why the same mistakes are made over and over. The authors have no answer. Yet many senior construction professionals are likely to respond, "The photographs are good to illustrate a point, but you will not find these conditions on my project." This would be contrary to the authors' experience. During their professional experience, which totals more than 70 years, the authors have managed and monitored hundreds of projects. The flaws discussed in this text have occurred repeatedly, regardless of the size or type of project. Every contractor can learn from the principles presented in this text.

1.6 Book Outline

This text is divided into four parts. Part I contains a chapter entitled "Weight of Expert Opinion." This chapter summarizes what other authors have said related to improving labor productivity. The remainder of the text discusses most of the issues detailed by these other authors. Improving cost and schedule performance begins before the bid is submitted with planning. Therefore, Part II contains two chapters dealing with planning. Chapter 3 describes a 17-step planning procedure. Chapter 4 explains a simple process for developing a site layout plan. Part III addresses site management factors that must be effectively handled if performance is to improve. In the seven chapters that form Part III, the authors discuss principles for mitigating the consequences of adverse weather, improving the efficiency of material management, efficient workforce management practices, effective activity and trade sequencing, avoiding congestion, and better subcontractor management. Part IV includes the special topics of schedule acceleration and environmental compliance. The last chapter of Part IV explains how to evaluate a project to assess where improvements can be made.

References

Sweet, J., and Schneier, M. M. (2009). *Legal aspects of architecture, engineering and the construction process,* 8th Ed., Cengage Learning, Stamford, CT.

CHAPTER 2

Weight of Expert Opinion

This chapter is organized according to the order of the remaining chapters in this text. The purpose of this chapter is to (1) summarize what various authors and practitioners have said about labor efficiency and (2) illustrate the relevancy of the remaining chapters in this text.

2.1 Planning

Considering the importance of construction planning, the literature is surprisingly void of published articles on effective planning. Most of the ASCE journal literature on planning seems to relate to environmental, transportation, and water resources planning; the owner planning process; and the planning topic of scope definition. There is little related to construction project planning, especially works that would be of use to contractors. Yet, developing a good plan is the first step toward maximizing profits. The literature topics related to planning that are reviewed are under the topical headings of (1) general, (2) processes and models, and (3) constructability.

General

As defined by the Construction Management Association of America (CMAA), "construction management is a professional service that applies effective management techniques to the planning, design, and construction of a project" (CMAA 2017).

There are numerous texts and articles on the critical path method (CPM) of scheduling, which describe planning as the process of defining the activities in a CPM schedule and determining the order of precedence of these activities (Hinze 2008). Planning as seen by CMAA and the authors is much more than this. The planning process as it relates to a CPM schedule is much too limited to define effective planning.

Laufer et al. (1993) describe the general planning process as the integration of a set of interdependent decisions that results in a plan. They described the plan as one that answers the following questions: (1) What is the involvement of the participating parties in planning? (2) What is the effort invested in planning? (3) How many types of plans are issued? and (4) What formats are used for plans?

They seem to suggest that planning is mostly a decision-making process, albeit a structured one. Their article is general and offers little or no specific guidance to contractors.

Processes and Models

Gibson et al. define preproject planning as the process of developing sufficient strategic information for owners to address risks and to decide whether to proceed with a capital project (1995). They point out that many construction industry experts believe early planning efforts have much influence on project success. Using a Level 1 IDEF () model, Gibson et al. define the steps in the owner preproject planning process as the following: (1) organize for preproject planning, (2) select alternatives, (3) develop project definition packages, and (4) make decisions. The major subprocesses of the owner preproject planning process were defined in more detail using Level 2 IDEF () models. Gibson et al.'s models make no assessment as to whether the models are complete or if they contribute positively to better cost and schedule performance; they merely report what owners currently do. The models are too general to be used as a framework for the contractor planning process. The models are strategic models used by industrial owners.

In a subsequent article by Gibson et al. (2006), they wrote of five major studies investigating preproject planning. The studies covered more than 200 capital projects (from 1994 to 2001), representing about $8.7 billion. Survey instruments and interviews were used to show that effective preplanning led to improved performance in terms of cost, schedule, and operational characteristics. The paper described a Project Definition Rating Index, which is a weighted matrix describing 70 scope definition elements. They conclude that (1) preplanning is a process that can positively affect capital project performance, (2) preplanning is a critical process that must be performed consistently on each project, (3) the project manager and team must ensure that they are performing the "right project," (4) the project manager and team must ensure that they are developing the "right work product" during preplanning, and (5) the project manager and team must choose the "right approach" to project design and construction execution.

In a 1997 article by Faniran et al., the authors present several strategies for improving contractor planning practices. Data and information were gathered via a questionnaire. Because a questionnaire was used, the article reports current practices, not necessarily best practices or fundamental principles. The authors conclude that planning would be more effective if there were (1) more investment of high-quality time in preconstruction planning, (2) less emphasis on developing schedules, and (3) more emphasis on developing operational plans. The writers made extensive use of statistical correlations and regression modeling as the basis of their conclusions, but it is hard to envision how the writers could support their conclusions because most correlation coefficients were below 0.50 and the best r_a^2 value was 0.60.

Constructability

There is little argument that design plays an important role in project performance. Most published articles related to design effects champion constructability reviews as a way to minimize design errors, identify more workable design solutions, and produce construction-friendly specifications, among other things.

An authoritative report on constructability reviews was written by Anderson and Fisher and published by the National Cooperative Highway Research Program (NCHRP) (Anderson and Fisher 1997). Anderson and Fisher observed that constructability reviews are often driven by project milestones and are conducted late in the design process as part of the final design review. Minimal reviews are done in the owner planning phase. In an NCHRP report by Thomas and Ellis, it was identified that the review may occur at a predetermined date, even if key documents needed for the review are not available (Thomas and Ellis 2001). Sadly, on most private, low-bid projects, there is little opportunity for contractor input into the design phase. The Anderson and Fisher report (1997) contains a good bibliography, and the subject of constructability is not discussed further in this book.

2.2 Site Layout

Most authors recognize the complexity of the site layout problem. All generally agree that there are usually multiple selection criteria and multiple constraints and that the plan changes over time. The site plan used at the beginning of the project is probably not suitable for the middle or later parts of the project.

For simplicity's sake, the most comprehensive algorithms use selection criteria to minimize travel distance or to minimize transportation cost (Mawdesley et al. 2002; Zouein and Tommelein 1999). The mathematical algorithms described in the various articles concentrate on positioning temporary facilities and locating laydown areas to satisfy constraints, while satisfying the minimum distance or cost objective. Although applying multiple criteria and addressing other features of the layout problem are perhaps possible, the site layout problem can quickly become quite complex. Furthermore, for a site where the site layout is especially important, travel distances and transportation costs may be of secondary importance. Relying upon the principles that ensure safety, adherence to the project schedule, and good labor productivity are of primary importance. The link between site layout and material management should not be lost. Yet none of the articles in the literature seem to adequately address this issue. Multiple objectives and numerous constraints are not easily expressed mathematically.

The published literature on developing a site layout is not helpful. The articles generally describe computerized, "black-box" solutions. Some approaches involve the development of an extensive knowledge base (Zouein and Tommelein 1999), whereas others do not rely on an extensive knowledge base (Mawdesley et al. 2002). Most algorithms relate to the positioning of temporary facilities and storage areas, whereas others are more limited, such as selecting the best location of cranes (Warszawski and Peled 1987).

2.3 Weather

There are two main literature sources that aid in understanding the origin of labor inefficiencies caused by adverse weather.

Framework

The first literature source is a report published by the Building Research Establishment (BRE) in the United Kingdom. This 1966 report was authored by Margaret Clapp. Based on field observations, the report describes how the loss of labor efficiency can occur in five categories. Three causes result when workers are being paid hourly wages. The first cause when workers are paid hourly wages is when a crew is forced to stop work and seek cover because of the severity of a weather event. The second cause results when a crew continues work during the event but the work is less efficient. The third cause of efficiency loss results from rework caused by unfavorable weather conditions.

Clapp also describes how labor efficiency losses can occur when workers are not being paid. The first cause in this category is absenteeism. This can result from extreme temperatures and humidity from a severe weather event. The second nonpaid cause is what Clapp calls loss of momentum. This occurs when a full work week cannot be worked because of a weather event. She argues that a shorter work week is less efficient than a normal work week.

The work of Clapp is consistent with the observations of the authors of this book. Her work, citing the five causes of inefficiency, provides a convenient framework for discussing labor inefficiencies caused by adverse weather conditions.

However, the adverse effect from weather events can last longer than the event itself. For example, rain can cause mud, which can affect an operation for days; snow may need to be shoveled for days after a storm; and weather enclosures may be damaged by high winds, affecting the work inside for days until repairs are made. Each of these examples has been observed repeatedly by the authors.

The second literature source is a book by Helander (1981). While he presented limited information, and his sources of information were not cited, he presented data to show that when exposed to extreme temperatures, crafts people are more adversely affected when they are engaged in cold weather work requiring fine motor skills compared with work requiring gross motor skills. The types of skills required were defined according to the trade. Boilermakers, carpenters, electricians, millwrights, pipefitters, surveyors, instrument technicians, and insulators use fine motor skills. Gross motor skills are required of laborers, bricklayers, ironworkers, teamsters, and operating engineers. While there is no scientific evidence to support such delineation, the framework posed by Helander seems reasonable and plausible.

Adverse weather conditions can occur from extremes in temperature and humidity or from weather events such as rain, snow, or high wind.

Temperature and Humidity

Three research publications are noteworthy in that the effects of temperature and humidity are described. Possibly the most widely cited source is a report published by the National Electrical Contractors Association (NECA 1974). This publication reports laboratory observations and measurements of two journeyman electricians who were confined in an environmental test chamber and were asked to install duplex receptacles on several mockup panels containing prewired outlet boxes. The temperature and humidity remained constant for two-hour periods. The length of the measured work period was one hour. The overall study lasted six days. After the study, efficiency loss curves were produced for hot to cold temperatures and for low to high humidity ranges. In general, the curves show that ideal temperatures were in the range of 40–80°F. Very low humidity had minimal effect on labor performance, but relative humidity above 60–70% was detrimental. High temperatures yielded greater efficiency losses than low temperatures, although the efficiency loss at 110°F was about the same as it was for –10°F. A wind chill factor was given, but this was not a central part of the study. No collaborative data were given to support the correctness of the wind chill factors.

The NECA study was based on laboratory data where fine motor skills were used. However, a study of masons by Grimm and Wagner (1974) was based on field data where gross motor skills were used. Masons were asked to construct straight walls (no corners or windows), and their output was measured. Ambient temperatures and humidity values were recorded. The study was conducted in Texas, so cold temperature observations were minimal. The Grimm study is one of the few studies of warm weather effects. Wind chill factors were not included in the study. The study showed that around 75°F was the ideal temperature, but when there were minor temperature variations, there was a significant reduction in efficiency, so much so that the cold weather aspects of the study are suspect. The cold weather conclusion of the Grimm study was that below 75°F there was a significant and rapid degradation in productivity. These findings are contradicted by the NECA study, which showed that for a 40-degree temperature range, the efficiency is essentially 100%. The findings from the NECA study are more consistent with field observations made by the authors than is the Grimm study.

Thomas and Yiakoumis reported the effect of temperature in an article based on field observations of concrete formwork, structural steel, and masonry (Thomas and Yiakoumis 1986). The zero loss of efficiency range was 50–60°F. Wind was not included in this study.

Two other published articles by Thomas et al. using field data show the effects of cold temperatures on the use of gross motor skills. In a 1999 article by Thomas et al., where structural steel erection was observed, efficiency losses were not measurable until the temperature fell below 20°F. The efficiency below 20°F was reduced to 50%.

In a second article, where congested work areas affected masonry work, the negative effect of cold temperatures was again not detected until the temperature fell below 20°F (Thomas et al. 2006). The efficiency below 20°F was reduced to 36%, or a loss of 64% on this project. On three of the six workdays where cold temperatures were a factor, cold temperatures delayed the start of mortar production by three to five hours.

All the articles by Thomas et al. show labor inefficiencies worse than those reported by Helander.

The effects in Fig. 2-1 and in Tables 2-1 and 2-2 are based on effects reported from the literature sources. The values shown are not arithmetic averages, as sometimes only a single number was reported. The values represent the authors' best estimate of the probable average effect, but individual project values vary widely.

Weather Events

Weather events include rain, snow, and high winds. The effects are thought of as lasting only one day; however, the negative effects can last for several days or more, for example, where rain leads to muddy conditions. The authors have observed the lingering effects of snow, rain, and high winds many times. Crews have lost a week or more of productive time because they had to shovel snow, pump water, deal with mud, and perform rework.

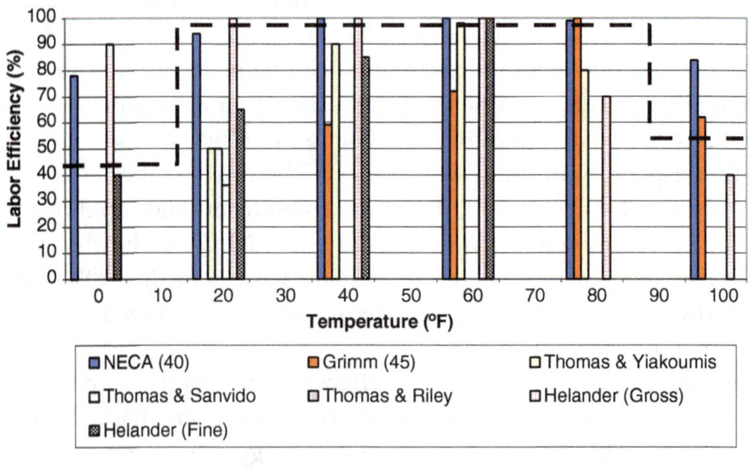

Fig. 2-1. Effect of Temperature on Labor Productivity

Table 2-1. Inefficiency Calculations

Project	Type Activity	Baseline Productivity	Actual Work Hours[a]	No. Workdays Affected	Inefficient Work Hours	Percentage of Normal Efficiency (%)	
						Rain	Snow
BT	Str. steel	1.50 wh/pc	189.5	4	67.3		55
RB	Str. steel	1.25 wh/pc	155	5	41.5		37
BJT	Masonry	0.082 wh/ft^2	62	2	14.0	29	
I99–9901	Formwork	0.066 wh/ft^2	70	2	53.0	317	
I99–9903	Formwork	0.085 wh/ft^2	110	7	62.8	133	
I99–9904	Formwork	0.083 wh/ft^2	120	2	51.0	74	
I99–9905	Formwork	0.098 wh/ft^2	120	1	107.3	848	
I99–9906	Formwork	0.141 wh/ft^2	206	3	102.7	99	
Total			626	26	391.0	132	46

Notes: BT is the Biotech building, RB is the Rider building, BJT is the Bryce Jordan tower, I99–xxxx are various bridges on Interstate 99, Str. steel means structural steel, wh means work hours, pc means piece.

[a]Work hour total is for only those workdays when a weather event occurred.

Table 2-2. Summary of Efficiency Losses Due to Weather

Cause of Labor Inefficiency	Estimated Percentage of Normal Daily Output
Cold temperatures (<20°F)	50
Hot temperatures (>85°F)	40
Moderate temperatures (20°F < T < 85°F)	None
Wind chill	Unknown
High winds	Unknown
Rain	60
Snow	50

Snow

In 1986, Thomas and Yiakoumis published an article quantifying weather effects derived from three projects (Thomas and Yiakoumis 1986). One project activity observed (identified as Project BT in Table 2-1) was structural steel. The work was affected by snow on four days. On those four days, the crew output was reduced to 55% of what would have normally been produced if the normal productivity had been achieved. Thus, the loss of efficiency was 45%. No lingering effects were observed.

On another project (Project RB in Table 2-1), also involving structural steel erection, the work was affected by snow and cold temperatures (Thomas et al. 1999). It snowed on five workdays, and the efficiency on these workdays was reduced to 37% of normal.

Rain

There have been a number of projects cited in the literature that were affected by rain. Table 2-1 contains pertinent data on these projects.

On Project BJT, rain affected masonry work on two days (Thomas et al. 2006). The labor inefficiency was calculated (see Table 2-1) as reduced to 29% of normal. The days when mortar operations were delayed were not counted.

Projects 9901–9906 involved bridge formwork on nine interstate highway bridges. The percentages of labor inefficiency caused by rain ranged from 74 to 848% of normal.

The inefficiency values show a wide range of susceptibility to precipitation, especially rain. The effect of high winds has not been determined.

Weather Summary

Except for the work of Clapp (1966) and the authors of this book, there has been little written about the effects of weather events on labor efficiency. Field data for eight projects were analyzed, and the pertinent data for all projects are given in

Table 2-1. Each of the activities on the eight projects reported in the table was fully exposed to the weather.

The combined loss of labor inefficiency for all projects was calculated for rain as 132% and snow as 46%, as shown in Table 2-1. The inefficient work hour percentage was calculated using the following equation:

$$\text{Labor Inefficiency } (\%) = \frac{\text{Inefficient Work Hours}}{\text{Total Work Hours} - \text{Inefficient Work Hours}} \times 1 \qquad (2\text{-}1)$$

Estimated loss of efficiency percentages are given in Table 2-2.

2.4 Material Management

There have been relatively few articles dealing with site material management practices. Work by Stukhart on bar coding was done to allow for easy identification and retrieval, but this work seems more applicable to larger industrial sites (Stukhart 1990).

Material management as defined herein involves storage, identification, retrieval, transport, and construction methods at the site. Each is indelibly linked to safety, productivity, and schedule performance. Thomas et al. show the effects on productivity of poor storage practices (1989). The difficulties in retrieval were readily obvious. In a later article, Thomas et al. also show the effect of delivery methods (1999). It was shown that the erection of structural steel directly from the delivery truck was the preferred method. Doing so also saved on-site storage costs and eliminated most double-handling. In essence, just-in-time material deliveries were preferred in this instance, but erecting directly from the delivery truck requires closer coordination with the fabricator to be successful.

In another article on site material management practices, Thomas and Sanvido show that effective site management practices can have a significant effect on schedule performance. The schedule slippage on the installation of permanent windows, precast panels, and duct ranged from 50% to 129% (2000).

Riley and Sanvido, Tommelein et al. and Thabet have described work space use and work area patterns in construction (Riley and Sanvido 1995, 1997; Tommelein et al. 1992; Thabet 1992). Riley and Sanvido use case studies to define work area patterns as being linear, random, horizontal, vertical, spiral, and building face (Riley and Sanvido 1995). They argue that space need patterns changed over time and that for effective use of resources, space needs must be predictable and rationally planned.

2.5 Workforce Management

The weight of expert opinion relative to workforce management is cited in five broad categories: (1) causes of loss of labor efficiency, (2) quantification of losses of labor efficiency, (3) getting the most out of the workforce, (4) workforce management practices, and (5) congestion and stacking of trades.

Causes of Loss of Labor Efficiency

The discussion of literature related to labor inefficiency is organized according to how the study was done: general, field or observational studies, and interviews and surveys. While labor productivity can be viewed as a complex issue, the literature suggests that there are relatively few factors that repeatedly cause disappointing performance outcomes.

General

A considerable body of knowledge exists on the causes of labor inefficiencies. Many articles and reports have been published in this domain. Four notable reports have been written by Sebestyen (United Nations 1965), Thomas and Smith (1990), Horner and Talhouni (1993), and Thomas et al. (2002). Each report contains extensive bibliographies of published literature on the causes and effects of labor inefficiencies, and the reader is referred to these reports for a comprehensive review of published literature. These reports make it clear that the main causes of labor inefficiencies are (1) ineffective utilization of resources (labor, materials, tools, equipment, and information), (2) unfavorable working conditions caused by management (congestion and out-of-sequence work), and (3) adverse weather.

The importance of management focus on the use of labor as a resource has only recently been quantitatively determined (Thomas et al. 2002). The authors found that on five bridge construction projects, more than half of the calculated inefficient work hours were the result of ineffective workforce management practices. This analysis led to further research, which showed that symbiotic crew relationships are more difficult to manage than are sequential relationships (Thomas et al. 2004).

Field or Observational Studies

Since 1980, the authors have observed and monitored labor productivity and performance on more than 200 active construction projects. The types of projects monitored include apartment buildings, high-rise condominiums, commercial offices, parking garages (aboveground and underground), government buildings, wastewater plants, highways, bridges, research and classroom buildings, indoor sports facilities, and residential sites. Activities studied in detail include masonry, concrete formwork, steel reinforcement, concrete placement, structural steel erection, caisson drilling, partition wall framing and drywall, duct installation, underground conduit work, excavation, asphalt paving, and concrete paving.

In the process of observing this work and documenting site conditions affecting the work, workforce management issues have been frequently noted

(Thomas and Horman 2006). Common workforce management deficiencies included insufficient work to perform, performing cleanup or incidental work in a sequential manner, overstaffing, and ineffective use of work teams. Each factor leads to more idle time.

As a result of these investigations, the authors have developed the Factor-Resource Model as a way to graphically illustrate the main influencing factors that adversely affect labor performance. This graphical input-output model was shown in Fig. 1-1. Its simplicity should be readily apparent. These are the root causes of inefficiencies and have been repeatedly observed on many projects by numerous authors. While there can be other factors, if one can manage the root causes as shown in Fig. 1-1, there is a high likelihood that the labor force will be productive.

Interviews and Surveys

A number of writers have sought input from industry professionals, contractors, and crafts workers through interviews, surveys, and questionnaires about the causes of lost labor efficiency. Numerous reports have been written. Even though the industry professionals, contractors, and crafts workers responded somewhat differently, there is much similarity in their responses, and many responses point to inadequate workforce management. Frequently cited factors leading to loss of labor efficiency are (1) lack of detailed planning, (2) lack of materials, (3) lack of tools, (4) rework, (5) congestion, and (6) poor workforce management practices.

Quantification of Losses of Labor Efficiency

While numerous articles quantify inefficiencies, the published literature in ASCE is surprisingly void of articles on quantification methodologies. However, one aspect is clear: labor inefficiencies are expensive. The costs of labor inefficiencies accrue rapidly. Where there is schedule acceleration, the cost overrun from inefficient labor hours (compared with the budgeted hours) can easily exceed 25–50%.

Articles and reports have been published describing the effects of inefficient labor hours caused by scheduled overtime (Thomas and Raynar 1997), poor material management and vendor relations (Thomas and Sanvido 2000), construction changes (Hanna et al. 1999a, b; Ibbs et al. 2007), disruptions (Thomas and Smith 1990; Thomas and Napolitan 1995; Horner and Talhouni 1993), and wasteful workforce management practices (Thomas et al. 2002).

Two articles specifically address quantification methodologies. Thomas and Volkman question the validity of the total cost method of quantification by arguing that there can be wide variations in work hour estimates from one city to another (2007). Thomas (2010) explains various difficulties in performing a measured mile analysis. He proposes a baseline productivity analysis that eliminates most of the deficiencies in a measured mile analysis.

Getting the Most out of the Workforce

The literature is sparse regarding efficient site management practices that focus on how to get the most out of the workforce.

Subassemblies and Modules (Methods)

An obvious labor-related practice is to use modules or preassemblies. It is common practice to preassemble reinforcement cages for columns and caissons. It is also often seen that crews assemble multiple pieces of ductwork at floor level before erecting the components as a single unit. A subassembly of three pieces of duct is illustrated in Fig. 2-2. It is common knowledge that in many instances, subassemblies offer significant efficiency advantages. Han and Thomas observed a 50% improvement in ductwork productivity where duct sections were connected into subassemblies like those shown in Fig. 2-2 (2002).

Multiskilling

Multiskilling has been proposed as a way to improve efficiency. Multiskilling is when one craftsperson performs multiple tasks ordinarily done by multiple union crafts Burleson et al. conducted a hypothetical study and report a potential labor savings of 5–20% and a potential 35% reduction in the required size of the workforce (1998). One obvious advantage of multiskilling is that alternate work assignments are easier, although some labor union agreements may limit this practice.

Workforce Management Practices

Previous research has identified several principles on which to base the development of construction sequences. Echeverry et al. (1991) define a collection of factors on which

Fig. 2-2. Three-Piece Subassembly of Duct, Redifer Commons, State College, PA

activity sequences are based, including physical dependencies, trade interactions, path interferences, and code regulations. Construction practitioners use these common dependencies every day to develop critical path method (CPM) schedules. More detailed planning that determines work locations and the flow of trades through building spaces is often needed to direct individual crews. Riley and Sanvido describe the key results of these plans to be building sequences (the order that floors are to be completed) and floor sequences (the order that work is completed in specific areas of each building floor) (1997). They discovered that projects with sequence problems resulted in more interferences for mechanical, electrical, and fire protection trades than all other trades combined. Basic sequencing benchmarks were identified for project sequences based on spatial interaction of structural, enclosure, rough-in, and finish phases. Horman et al. describe sequence plans (2006). Alvanchi et al. introduce the notion that working hours can affect performance (2012). They also discuss the effect of the length of work and rest periods, time of day, and amount of overtime.

Successful practices and space planning techniques were organized into a set of steps to help develop an efficient sequence plan for the mechanical, electrical, and fire protection trades. These steps are the following:

Subdivide the floor plan into work zones. Suggested zones that may be considered independently are the core, perimeter, and areas in between. Large floor areas may be subdivided into halves or quarters. Another scheme may be to divide the floor into the corridor and rooms.

Determine the order in which the floors will be completed. A top-down sequence is often effective for interior and finish trades on low-rise buildings.

Determine the order of completion for each floor. Each zone on a floor may have its own sequence.

Specify and enforce the zonal sequence for each floor.

Determine the work rate for each zone based on the driving activity in that zone. Often there is only one task that establishes the pace of a larger grouping of tasks. This is referred to as the driving activity. Trades should size crews to meet target work rates. Additional flexible crews or teams can support frontline crews as needed. Target work rates should be established before negotiations with subcontractors.

The authors have published several articles that provide steps contractors can take that will improve site construction management. These articles have been written in the topical areas of planning (Thomas and Ellis 2007), weather mitigation (Thomas and Ellis 2009), workforce management (Thomas and Horman 2006), material management (Thomas et al. 2005), congestion (Thomas et al. 2006), and subcontractor management (Thomas 2011). Riley and Sanvido have published similar writings related to sequencing (1997).

Congestion and Stacking of Trades

The literature that addresses work space congestion largely focuses on quantifying the effects of trade stacking for the purpose of a claim presentation. There is limited information in the existing literature on how to avoid congestion, and no quantitative research on the relationship between planning and work space congestion.

Techniques to avoid congestion among multiple trades are characterized as planning activities that account for the use of space by trades for work spaces, material storage, and equipment. Riley and Sanvido (1995) characterize 12 unique characteristics of construction activities that require space and a planning technique that considers the patterns of space requirements needed by different types of construction work. This research also characterized key sequencing factors for construction trades, including the definition of work flow in terms of work rate and direction to minimize interferences among trades (Riley and Sanvido 1997). A technique to analyze the progress of trades through multiple work zones and assess potential interferences was developed by Thabet (1992).

The literature generally follows an approach in which labor performance is related to the area of work space available per worker. This approach appears to be more suited to manufacturing settings, where workstations are stationary, and for broad characterizations of construction projects. Because of the fluid nature of construction operations, a gross area per worker methodology must be calculated as an average over an extended period of time. This approach can grossly understate the effect and make congestion seem like a minor problem. This method also ignores the possibility that there may be causes of congestion other than too many workers, for example, cramped work spaces and stored materials that impede the work. Nevertheless, it is useful to report the findings of other writers in construction settings.

An informal presentation conducted by people working for the Mobile Oil Co. was based on the density of workers on refinery construction. Their unpublished study reports that subcontractors needed more area per worker than did direct-hire workers. For maximum efficiency for subcontractors, each worker needed an average of 250 ft^2, whereas a direct-hire worker needed 200 ft^2.

Smith (1987) reports losses of productivity similar to those in the Mobil Oil study. Based on offshore work on oil drilling platforms, he reports that maximum productivity occurred when workers had at least 320 ft^2 (99 m^2) per person.

Logcher and Collins (1978) studied the setting of floor tile on five projects in New York and Boston. They found that while more open area allowed more freedom of movement, the productivity was affected only slightly by changes in the floor area to perimeter ratio. In their study, the productivity was unrelated to the square feet per person, although on all projects, the area per person exceeded 300 ft^2.

Horner and Talhouni surveyed the literature and published their findings in a report to the Chartered Institute of Building in the United Kingdom (1993). They state that the data in this topical area (congestion) are especially sparse. Thomas and Smith (1990) report similar findings, leading one to conclude that there has been little scientific research done on the effect on labor productivity caused by congested work areas. The consensus from these reports is that for maximum efficiency, workers need at least 250–300 ft^2 per person.

2.6 Activity and Trade Sequencing

Research into sequencing is plentiful but not particularly helpful. The information is organized in the following categories:

Planning

- Sequence plans and execution,
- Successful methods of dealing with unknowns,
- Reliability of the planning effort coordination,
- Successful coordination processes and methods, and
- Flow of coordination information.

Materials and Methods

- Examples of materials and methods that create flow and
- The role of material handling in sequencing.

Expectations

- Owner expectations of sequence planning and
- Realistic expectations for general contractors.

While the issue of sequencing may be considered a project management function, this broad topic is rarely discussed in the context of production. Instead, existing project management practices are directed toward tracking progress. The following sequencing issues are evident on many projects:

- The relationship between trade sequencing and project management in many organizations is not well understood or defined. There is a great need for production management or tactical planning. The focus in many organizations is on scheduling or strategic planning.
- Gaining a competitive advantage and generating cost savings by improved sequencing practices is a key method to improving profitability.
- There is a widely perceived inability or unwillingness of general contractors and construction managers to implement detailed sequence plans.
- The perpetual acceptance of interference problems and out-of-sequence work as an inevitable aspect of building construction leads to complacency and isolation during the planning process when proactive and team thinking is needed.

2.7 Subcontractor Management

How a contractor manages subcontractors is an important component of construction project management because many project types are constructed largely by subcontractors (Thomas 2011). Yet the published literature is surprisingly void of guidance on what constitutes effective management of subcontractors.

Congestion

The literature that addresses workspace congestion largely focuses on quantifying the effects of trade stacking for the purpose of a claim presentation. There is limited information in the existing literature on how to avoid congestion, and no quantitative research on the relationship between planning and workspace congestion.

Techniques to avoid congestion among multiple trades are characterized as planning activities that account for the use of space by trades for workspaces, material storage, and equipment. Riley and Sanvido (1996) characterize 12 unique characteristics of construction activities that require space, and a planning technique that considers the patterns of space requirements needed by different types of construction work (Riley and Sanvido 1997a). This research also characterized key sequencing factors for construction trades, including the definition of work flow in terms of work rate and direction to minimize interferences among trades (Riley and Sanvido 1997b). A technique to analyze the progress of trades through multiple work zones and assess potential interferences was developed by Thabet (1992).

The quantification literature generally follows an approach in which labor performance is related to the area of workspace available per craftsman. This approach appears to be more suited to manufacturing settings, where workstations are stationary, and for very broad characterizations of construction projects. Because of the fluid nature of construction operations, a gross area per craftsman methodology must be calculated as an average over an extended period of time. This approach can grossly understate the impact and make congestion seem like a minor problem. This method also ignores the possibility that there may be causes of congestion other than too many craftsmen, for example, cramped workspaces and stored materials that impede the work. Nevertheless, it is useful to report the findings of other writers in construction settings.

An informal presentation conducted by persons working for the Mobile Oil Co. was based on the density of workers on refinery construction. Their unpublished study reports that subcontractors needed more area per craftsmen than did direct hire craftsmen. For maximum efficiency for subcontractors, each craftsman needed an average of 200 ft^2 whereas a direct hire craftsman needed 250 ft^2.

Smith (1987) reports losses of productivity similar to those in the Mobil Oil study. Based on offshore work on oil drilling platforms, he reports that maximum productivity occurred when craftsmen had at least 320 ft^2 (99 m^2) per person.

Logcher and Collins (1978) studied the setting of floor tile on five projects in New York and Boston. He found that while more open area allowed more freedom of movement, the productivity was affected only slightly by changes in the floor area to perimeter ratio. In his study, the productivity was unrelated to the ft^2/person, although on all projects, the area per person exceeded 300 ft^2.

Horner and Talhouni surveyed the literature and published his findings in a report to the Chartered Institute of Building (1993). They state that the data in this area (congestion) are especially sparse. Thomas (1990) reports similar findings leading one to conclude that there has been very little scientific research done on the impact on labor productivity of congested work areas. The consensus from the Horner and Thomas reports is that for maximum efficiency, craftsmen need at least 250–300 ft^2 per person.

2.8 Schedule Acceleration

The literature is not helpful relative to schedule acceleration. Most articles on acceleration have been written by attorneys and consultants describing legal theories and documentation practices so that the contractor can file a successful claim at the conclusion of the project. They offer little or no advice on how a contractor is to avoid acceleration, how to recognize acceleration, or how to manage acceleration.

Only one helpful ASCE article was identified. That article showed significant variations in the weekly workload. It presented an inefficient labor quantification methodology based on variations in workload. The methodology included labor inefficiencies from scheduled overtime and overstaffing (Thomas 1999b).

A series of reports for the Electrical Contracting Foundation document how to assess contract risks (Thomas 1998), how to recognize that acceleration is coming (Thomas 2002), what to do to avoid some of the negative consequences, how to quantify labor inefficiencies (Thomas and Oloufa 1996), and strategies for negotiating loss of labor efficiency claims (Thomas 1999a).

Although the acceleration literature is sparse, all authors agree that acceleration is a bad situation; it cost lots of money, and the project participants often end up in court.

2.9 Lean Construction Principles

The Lean Construction Institute has championed some principles that are designed to promote better labor performance. These principles tend to be general rather than specific and may not be applicable to all situations.

Ballard and Howell (1998) proposed the following:

- Halt work to correct systematic problems—Perform work only when it is free of design defects.
- Pulling versus pushing—Construction, not design information, should drive resource needs.
- One-piece flow—Crews and teams should complete all the work possible on each pass through the area.
- Synchronize and align—Get crews to work at similar paces.

- Transparency—Share progress information, feedback, and challenges with everyone.

Supply chain management has a direct effect on project sequencing, because available materials and services must be synchronized for efficient field production. Lean supply recommendations for specialty trades include the following:

- Make suppliers into business partners.
- Give specialty contractors design-build responsibility.
- Reimburse costs to get control over production processes.
- Integrate planning and control systems.

Lean performance methods are used to shape sequence guidelines, with particular emphasis on those that relate to methods and materials. Lean performance improvement strategies for construction (Ballard and Howell 1998) include the following:

- Minimize the effect of erratic resource delivery and quality.
- Stabilize work processes by maintaining workable backlogs and assigning only from backlogs.
    - Eliminate delays for materials, equipment, and crew interference.
    - Make the supervisor a manager and require detailed planning to be done.
    - Get the facts about delays and rework.
    - Reduce costs.
    - Benefit from constructability.
    - Structure on-site work as close as possible to workshop conditions.
    - Develop workers with multicraft capabilities.
    - Maintain tools and equipment for zero downtime.
- Reduce rework by completing work the first time, identifying and learning from errors, and tracking repetitive errors to their causes.
    - Reduce the number of organizational errors.
    - Reduce the duration of work.
- Advance mobilization with the constraints of available backlog and reliable work flow.
    - Reduce backlog quantities to reflect more reliable deliveries.
- Divide construction sites into subprojects and facilitate functional team approaches.

2.10 Summary of Weight of Expert Opinion

If one examines the published literature, it is obvious that there are many opportunities to increase profits through better site management. However, there is little research devoted to this topic, especially as it relates to the important topics of subcontractor management and schedule acceleration. Instead, researchers tend to focus on quantification issues, intricate mathematical models, or manufacturing practices that may not be applicable to construction. The larger issue of what to do to increase profitability remains unanswered.

References

Alvanchi, A., Lee, S., and AbouRizk, S. (2012). "Dynamics of working hours in construction." *J. Constr. Eng. Manage.*, 66–77.

Anderson, S. D., and Fisher, D. J. (1997). "Constructibility review process for transportation facilities." NCHRP Rep. No. 390, National Cooperative Highway Research Program, Transportation Research Board, Washington, DC.

Ballard, G., and Howell, G. (1998). "Shielding production: An essential step in production control." *J. Constr. Eng. Manage.*, **124**(1), 11–17.

Burleson, R. C., Haas, C. T., Tucker, R. L., and Stanley, A. (1998). "Multiskilled labor utilization strategies in construction." *J. Constr. Eng. Manage.*, **124**(6), 480–489.

Clapp, M. A. (1966). "The effect of adverse weather conditions on five building sites." Construction Current Paper No. 21, Building Research Establishment, Watford, U.K., 171–180.

CMAA (Construction Management Association of America). (2017). ⟨http://cmaanet.org/cm_is.php⟩

Echeverry, D. C., Ibbs, C. W., and Kim, S. (1991). "Sequencing knowledge for construction scheduling." *J. Constr. Eng. Manage.*, **117**, 118–130.

Faniran, O. O., Oluwoye, J. O., and Lenard, D. J. (1998). "Interactions between construction planning and influence factors." *J. Constr. Eng. Manage.*, **124**(4), 245–256.

Gibson, G. E., Jr., Kaczmarowski, J. H., and Lore, H. E., Jr. (1995). "Preproject-planning process for capital facilities." *J. Constr. Eng. Manage.*, **121**(3), 312–318.

Gibson, G. E., Jr., Wang, Y. R., Cho, C. S., and Pappas, M. P. (2006). "What is preproject planning anyway?" *J. Constr. Eng. Manage.*, **22**(1), 35–42.

Grimm, C. T., and Wagner, N. K. (1974). "Weather effects on mason productivity." *J. Constr. Div.*, **100**(3), 319–335.

Han, S. Y., and Thomas, H. R. (2002). "Quantification of labor inefficiency." *Proc., 10th Triennial Symp. of Int. Council on Innovation in Building Construction*, Cincinnati, OH.

Hanna, A. S., Russell, J. S., Gotzion, T. W., and Nordheim, E. V. (1999a). "Impact of change orders on labor efficiency for mechanical construction." *J. Constr. Eng. Manage.*, **125**(3), 176–184.

Hanna, A. S., Russell, J. S., Nordheim, E. V., and Bruggink, M. J. (1999b). "Impact of change orders on labor efficiency for electrical construction." *J. Constr. Eng. Manage.*, **125**(4), 224–232.

Helander, M., ed. (1981). *Human factors/ergonomics for building and construction*, Wiley, New York.

Hinze, J. W. (2008). *Construction planning and scheduling*, Pearson Prentice Hall, Upper Saddle River, NJ.

Horman, M. J., Crosz, M. R., and Riley, D. R. (2006). "Sequence plans for electrical construction." *J. Constr. Eng. Manage.*, **132**, 363–372.

Horner, R. M. W., and Talhouni, B. T. (1993). *Effects of accelerated working, delays and disruptions on labour productivity*, Chartered Institute of Building, London, 40.

Ibbs, C. W., Nguyen, L. D., and Lee, S. (2007). "Quantified impacts of project change." *J. Prof. Issues Eng. Educ. Pract.*, **133**(1), 45–52.

Laufer, A., Shapira, A., Cohenca-Zall, D., and Howell, G. A. (1993). "Prebid and preconstruction planning process." *J. Constr. Eng. Manage.*, **119**(3), 426–444.

Logcher, R. D., and Collins, W. C. (1978). "Management impacts on labor productivity." *J. Constr. Eng. Manage.*, **104**(4), 447–41.

Mawdesley, M. J., Al-Jibouri, S. H., and Yang, H. (2002). "Generic algorithms for construction site layout in project planning." *J. Constr. Eng. Manage.*, **129**(5), 418–426.

NECA (National Electrical Contractors Association). (1974). *The effect of temperature on productivity*, Bethesda, MD, 28.

Riley, D. R., and Sanvido, V. E. (1995). "Patterns of construction-space use in multistory buildings." *J. Constr. Eng. Manage.*, **121**(4), 464–473.

Riley, D. R., and Sanvido, V. E. (1997a). "Patterns of construction space use in multistory buildings." *J. Constr. Eng. Manage.* **121**(4), 464–473.

Riley, D. R., and Sanvido, V. E. (1997b). "Space planning method for multistory building construction." *J. Constr. Eng. Manage.*, **123**(2), 171–180.

Smith, A. G. (1987). "Increasing onsite production." *AACE Trans.*, 4.1–4.13.

Stukhart, G. (1990). "Bar-code standardization in industrial construction." *J. Constr. Eng. Manage.*, **116**(3), 416–431.

Thabet, W. A. (1992). "Space constrained resource constrained schedule system for multistory buildings." Ph.D. dissertation, Dept. of Civil Engineering, Virginia Polytechnic Institute and State Univ., Blacksburg, VA.

Thomas, H. R. (1998). "Fundamentals of contract risk management for electrical contractors." Electrical Contracting Foundation, Bethesda, MD, 70.

Thomas, H. R. (1999a). "Negotiating loss of labor efficiency claims." Electrical Contracting Foundation, Bethesda, MD, 85.

Thomas, H. R. (1999b). "Schedule acceleration, workflow, and labor productivity." *J. Constr. Eng. Manage.*, **126**(4), 261–267.

Thomas, H. R. (2002). "Early warning signs of project distress." PTI Rep. No. 2002-34, The Electrical Contracting Foundation, Bethesda, MD, 45.

Thomas, H. R. (2010). "Quantification of losses of labor efficiency: Innovations in and improvements to the measured mile." *J. Legal Affairs Dispute Resolut. Des. Constr.*, **2**, 106–112.

Thomas, H. R. (2011). "Fundamental principles of subcontractor management." *Pract. Period. Struct. Des. Constr.*, **16**(3), 106–111.

Thomas, H. R., and Ellis, R. D., Jr. (2001). "Avoiding delays during the construction phase of highway projects." NCHRP Rep. on Project No. 20-24, National Cooperative Highway Research Program, Transportation Research Board, Washington, DC, 77.

Thomas, H. R., and Ellis, R. D., Jr. (2007). "Contractor prebid planning principles." *J. Constr. Eng. Manage.*, **133**, 542–552.

Thomas, H. R., and Ellis, R. D., Jr. (2009). "Fundamental principles of weather mitigation." *Pract. Period. Struct. Des. Constr.*, **14**, 29–35.

Thomas, H. R., and Horman, M. J. (2006). "Fundamental principles of workforce management." *J. Constr. Eng. Manage.*, **132**, 97–104.

Thomas, H. R., M. J. Horman, and U. E. L. de Souza. (2004). Symbiotic Crew Relationships and Labor Flow. *J. Constr. Engr. and Mgmt.*, **130**(6), 908–917.

Thomas, H. R., Horman, M. J., de Souza, U. E. L., and Zavrski, I. (2002). "Benchmarking of labor-intensive construction activities: Lean construction and fundamental principles of workforce management." Rep. 276, The International Council for Research and Innovation in Building and Construction (CIB), Rotterdam, Netherlands, 156.

Thomas, H. R., and Napolitan, C. L. (1995). "Quantitative effects of construction changes on labor productivity." *J. Constr. Eng. Manage.*, **121**(3), 290–296.

Thomas, H. R., and Oloufa, A. A. (1996). "Strategies for minimizing the economic consequences of schedule acceleration and compression." PTI Rep. No. 9615, Electrical Contracting Foundation, Bethesda, MD, 45.

Thomas, H. R., and Raynar, K. A. (1997). "Scheduled overtime and labor productivity: Quantitative analysis." *J. Constr. Eng. Manage.*, **121**(3), 181–188.

Thomas, H. R., Riley, D. R., and Messner, J. I. (2005). "Fundamental principles of site material management." *J. Constr. Eng. Manage.*, **131**, 808–815.

Thomas, H. R., Riley, D. R., and Sanvido, V. E. (1999). "Loss of labor productivity due to delivery methods and weather." *J. Constr. Eng. Manage.*, **125**, 39–46.

Thomas, H. R., Riley, D. R., and Sinha, S. K. (2006). "Fundamental principles for avoiding congested work areas on masonry work—A case study." *Pract. Period. Des. Constr.*, **11**, 197–205.

Thomas, H. R., and Sanvido, V. E. (2000). "The role of the fabricator in labor productivity." *J. Constr. Eng. Manage.*, **126**(5), 358–365.

Thomas, H. R., Sanvido, V. E., and Sanders, S. R. (1989). "Impact of material management on productivity—A case study." *J. Constr. Eng. Manage.*, **115**(3), 370–384.

Thomas, H. R., and Smith, G. R. (1990). "Loss of construction labor productivity due to inefficiencies and disruptions: The weight of expert opinion." Report to the National Science Foundation, PTI Rep. No. 9019, Pennsylvania State Univ., Pennsylvania Transportation Institute, University Park, PA, 181.

Thomas, H. R., and Volkman, R. C. (2007). "How reliable is the modified total cost or total cost method?" *J. Prof. Issues Eng. Educ. Pract.*, **133**(1), 74–77.

Thomas, H. R., and Yiakoumis, I. (1986). "Factor model of construction productivity." *J. Constr. Eng. Manage.*, **133**(1), 623–639.

Tommelein, I. D., Castillo, J. D., and Zouelin, P. P. (1992). "Space-time characterization for resource management on construction sites." *Proc., 8th Conf. on Computing in Civil Engineering*, ASCE, New York, 623–630.

United Nations Committee on Housing, Building, and Planning. (1965). *Effect of repetition on building operations and processes on site*, New York.

Warszawski, A., and Peled, N. (1987). "An expert system for crane selection and location." *Proc., 4th Int. Symp. on Robotics and Artificial Intelligence in Building Construction*, Vol. **1**, Israel Institute of Technology and Building Research Station–Technion, Haifa, Israel, 64–68.

Zouein, P. P., and Tommelein, I. D. (1999). "Dynamic layout planning using a hybrid incremental solution method." *J. Constr. Eng. Manage.*, **125**(6), 400–408.

PART II

Planning

CHAPTER 3

Fundamental Principles of Planning

Planning is a cornerstone of effective construction project management and improving performance, yet there exists little or no framework or guidance on how to do effective planning (Thomas and Ellis 2007). Most published discussions on planning are presented in conjunction with the development of a critical path method (CPM) schedule. These discussions of planning are limited to the presentation of text material defining activities or establishing precedence relationships (Hinze 2008). The questions of what steps should be taken, what should be done, and how one should plan remain unanswered. In this chapter, a 17-step procedure is described in which a holistic view of planning is presented, from risk analysis to the development of submittal schedules. If this procedure is followed, the result will be lower costs, shorter schedules, and increased profits.

3.1 Risk-Based Planning

Relative to planning, project risks can be organized into the following four categories: (1) things that can go wrong, (2) surprises (or the unknown), (3) contractual risks, and (4) relations with project team members. These risks are shown conceptually in Fig. 3-1. There is no relationship between the categories listed above and the proportions shown in Fig. 3-1. The focus of planning is to develop plans to avoid the risks or to minimize the economic effects from events in these four categories. Two situations commonly occur on a construction site: things go wrong, and there are surprises. One can anticipate and plan for many of the things that can go wrong. One can also plan for some surprises. Other elements of risks include contractual risks and relations with project team members. Some planning can be done for these two areas of risk. Where the risks are foreseeably high, the contractor may choose not to submit a bid. Through a risk-based planning process, a contractor can plan for and negate many project risks, or if the project is too risky, decline to submit a bid. Negating risks leads to a more profitable construction project.

This chapter defines the short-range (as opposed to strategic) steps that make up an effective contractor risk-based project planning process for small- and medium-size projects. Because many larger projects can be viewed as a collection of smaller

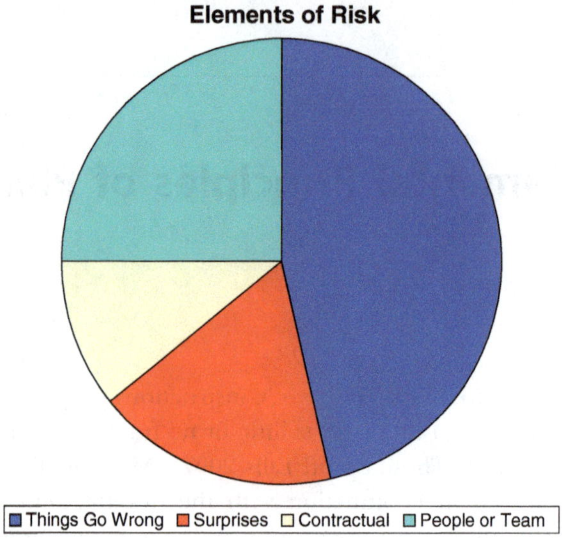

Fig. 3-1. Broad Categories of Project Risk

projects, there is application to larger projects as well. Following these steps will lead to better labor productivity, lower cost, and shorter schedules. Thus, the objective of risk-based planning is to anticipate surprises and things that may go wrong and to negate or eliminate these risks through better planning. Thus, it is important for a contractor to answer the question: What can go wrong?

3.2 Steps in Risk-Based Planning

The general approach to planning involves two broad steps: (1) deciding on preliminary considerations and (2) developing detailed plans. The preliminary considerations include a risk assessment to decide whether to submit a bid. If a decision is made to submit a bid, bid preparation involves deciding on a preliminary plan followed by optimizing the preliminary plan through the preparation of various detailed plans (step 2). Other preliminary considerations refer to preparing a preliminary schedule; developing a contracting strategy and a general strategy; selecting major means, methods, and equipment; and revising the preliminary schedule. Once the preliminary considerations have been finalized, various detailed plans need to be developed. Detailed plans include material delivery schemes, site layout plans, erosion and sediment control (E&S) plans, environmental considerations, CPM analyses, sequences that are essential to success (ETS sequences), and operational plans. Because these are specific to the project and the techniques used may vary. The techniques used to develop detailed plans are described in lieu of describing the plans themselves. The techniques described are the short interval production schedule (SIPS), velocity (production rate) charts, and linear scheduling. A critical path method (CPM) schedule is another technique that is frequently used, but it is not discussed herein because there are numerous texts that are devoted exclusively to CPM schedules (Hinze 2008). Other techniques include sequence plans and coordination drawings. Many of the detailed aspects of planning and executing and for formulating low-cost plans are covered in subsequent chapters. The goal is to construct the project in as short a time as practical while minimizing the need for field labor. After the notice of award is given, the planning function does not cease. Contractors must plan for submittals and managing changes.

3.3 Case Study—State College Municipal Building

The steps in the planning process are best illustrated with a case study project where the lack of planning was readily obvious. The steps are applied retrospectively to this project. The case study project is the three-story State College Municipal Building (SCMB). It was constructed in State College, Pennsylvania, in 2002 at a cost of about $5 million.

The building superstructure is structural steel, and the floor slabs are precast concrete. The project also contains a full-footprint basement. Construction of the basement required excavation, rock removal, footers, and the construction of a reinforced concrete wall. The building facade is masonry. An auditorium is included on the top floor. Fig. 3-2 shows the status of the SCMB construction after the precast slabs were installed.

The construction site was small and was constrained on all four sides by city streets and existing buildings. Ingress and egress points to the site were limited.

Fig. 3-2. State College Municipal Building (SCMB) during Construction

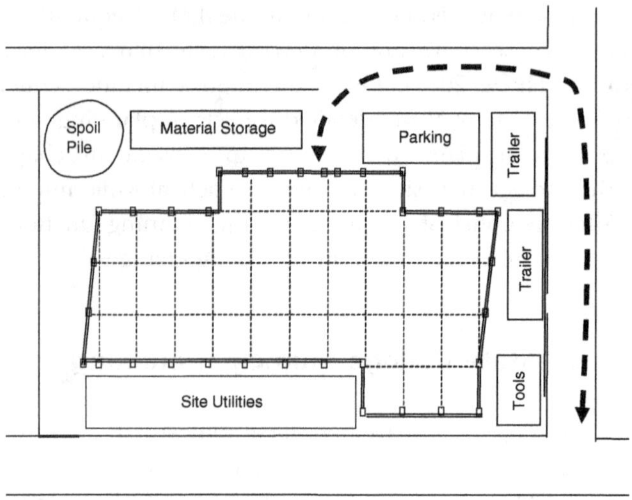

Fig. 3-3. Actual Site Layout for SCMB

The building footprint took about 60–70% of the site area, so space for on-site storage was minimal. The constrained nature of the site is shown in Fig. 3-3, which is the site layout used by the contractor throughout the project. The contractor used only one ingress and egress point for material deliveries. Using this layout, delivery trucks had difficulty accessing the site, and drive-through deliveries were not possible. The contractor chose to allocate space for an earthen spoil pile, trailers, and a parking area within the site boundary. Trailers and the parking area were located first. Material storage and other needs were assigned to the remaining unoccupied space.

The construction of this project was challenging, even though it was a relatively small project. Careful planning was justified.

3.4 Paramount Determination

For all projects during the bid period and before beginning to prepare the estimate, a paramount question to answer is this: Using conventional means and methods, can the contractor complete the project within the time allotted by the contract? If not, unconventional means and methods may need to be applied. Because time is money, the goal of contractors should always be to complete the project in as short a time as practical, even if adequate time is provided for in the contract. Shorter schedules minimize the contractor's risk exposure. It has been observed by the authors that contractors tend to use all the time allowed by the contract, even if a shorter schedule is practical. Some of the planning steps outlined in the following may be deferred on some projects until after the intent to award is given, but all steps should be done before project execution begins.

3.5 General Approach to Planning

Effective planning requires (1) assessment of risks, (2) settling on preliminary strategies, (3) considerations that reflect the risk assessments, and (4) making detailed plans that incorporate the assessments and strategies. All functions are linked. The assessment of risks includes the following:

- Things that can go wrong;
- Surprises;
- Contract risk;
- Other (e.g., people or payments); and
- Can I finish on time?

A risk assessment should always be done before the decision to submit a bid is made. Contractors may choose to avoid bidding on projects where there are high risks. Once the decision to bid is made, preliminary considerations must be decided upon. These include the following:

- Strategies and preliminary considerations;
- A preliminary execution plan (schedule);
- Contractual strategy (Which parts should be subcontracted?);
- General strategy (Where should I begin and how should I proceed?);
- Unconventional means and methods (e.g., multiple workstations);
- Construction equipment; and
- A revised execution schedule.

Once strategies and considerations have been decided upon, various plans need to be developed that describe in detail how the work will be done. Following these plans will minimize risks. These plans include the following:

- Detailed plans:
    - Material delivery schemes,
    - Site layouts, and
    - Weather mitigation plans,
- Erosion and sediment control (E&S) plans,
- Additional drainage plans,
- Footprint drainage,
- Drainage of the rest of the site,
- Runoff from off site onto the site,
- Building enclosure and sealing,
- Environmental and regulatory considerations,
- CPM analyses and time windows,
- Sequences that are essential to success (ETS sequences),

- Operational plans,
- Floor or area completion,
- Material delivery plans,
- Material distribution plans,
- Sequence plans,
- Coordination drawings,
- Final execution schedule,
- Submittal and shop drawing schedule, and
- Change management.

Of course, other risk assessments, preliminary strategies and considerations, and plans may be justified.

To illustrate the relationship between risk assessments, preliminary considerations, and detailed plans, consider an example. Suppose that a risk assessment has been made that by using conventional means and methods, there is a high probability that concrete placement will be delayed until the winter months. This poses a significant risk to the contractor's operation. Nevertheless, the contractor has decided to submit a bid.

To expedite the placement operation, the contractor adopts the following preliminary consideration and incorporates any added cost into the bid: Use several pump trucks at multiple workstations and apply concurrent scheduling practices. These strategies will minimize the risk of winter placements. Next, during the planning process, he or she develops detailed plans to ensure that the strategies are effectively implemented.

3.6 The Project Planning Process

The process described here should be viewed as a starting point, and more steps can be added as the need arises. However, the process described is generalized so as to be applicable to numerous projects. The steps in the planning process include assessments, strategies, and plans. The planning process is shown in Fig. 3-4. The steps are the following: (1) assess risks, (2) develop a preliminary execution plan, (3) decide on key macro strategies, (4) determine equipment needs and methods, (5) revise the preliminary execution plan, (6) decide on material delivery schemes, (7) develop E&S and site drainage plans, (8) understand environmental and regulatory issues, (9) develop site layout plans, (10) identify the sequences that are essential to success (ETS), (11) calculate time windows, (12) develop detailed operational plans, (13) develop proactive strategies to ensure construction input into design, (14) finalize the execution plan (schedule), (15) communicate and enforce plans, (16) develop a submittal and shop drawing schedule, and (17) plan for change management. Because the process should be flexible and each step should provide input to the next step, it is permissible to revise prior step(s) if a better plan emerges.

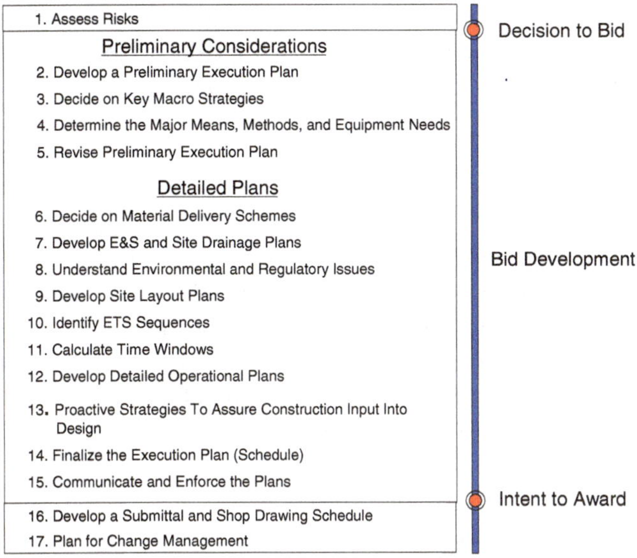

Fig. 3-4. The Planning Process

Assess Risks

Contractual Risks—The discussion on planning begins with an assessment of the contract risks. Some contracts are known to contain harsh language that imposes significant risks on contractors and subcontractors (Thomas et al. 2000). These risks need to be known before a decision to bid is made. Additionally, contracts may contain language related to means and methods, when certain operations can be performed, liquidated damages, intermediate milestone dates, and much more. A careful reading of the contract is paramount. Pay special attention to the termination clause, payment provisions, and the acceleration clause.

Noncontractual Risks—Risks go beyond what is written in the contract. One noncontractual risk that should be evaluated is the likelihood of a time extension, irrespective of the contract language. This aspect is often a function of the type of project. For instance, for schools and sports stadiums, extensions of time are unlikely. If a wastewater plant is under a court order to upgrade its treatment processes, the owner will be reluctant to grant an extension of time, regardless of what the contract says. These situations increase the contractor's risk and are especially problematic if the start of construction is delayed.

The reputation of the key players and organizations is another important risk consideration that needs to be evaluated. Some people are just hard to get along with. Others are sticklers for details. Will the owner pay on time? Do the contract documents appear complete and generally free from errors and omissions? Will the parties make timely decisions? These are but a few considerations that can make a project risky for the contractor.

A contractor should know the risks in a project before making a decision to submit a bid. One good rule of thumb is to leave risky projects to your competitors. There is little, if any, profit to be made on a risky project. After the risks are known, an educated decision can be made on whether to submit a bid.

Example

How an owner handles requests for an extension of time on a construction project is filled with risks. The court case of *Alexander & Shankle, Inc., v. Metropolitan Government of Nashville and Davidson County* (A&S) sheds some light on contractor risks and owner obligations (*Alexander & Shankle, Inc., v. Metropolitan Government of Nashville and Davidson County* 2007). Some of the more notable risks that materialized and adversely affected A&S were the following:

- Delays at the beginning of the project,
- Owner tardiness in decision making,
- Termination,
- Winter masonry work, and
- Scope changes.

If the contractor had done a diligent risk assessment, some or all of these risks could have been foreseen, and the decision to submit a bid might have been different. The factual aspects of the case are given below.

The Project

On March 25, 2003, the Metropolitan Government of Nashville and Davidson County (Metro) entered into a contract with Alexander & Shankle, Inc. (A&S), under which A&S became general contractor for the construction of Oliver Middle School and Shayne Elementary School, which were both to be constructed on the same site. The A&S bid was $11,841,000. A&S experienced numerous delays; however, Metro refused to extend the substantial completion date of the project.

The Contract

The contract provided the following:

> [a]ll limitations of time set forth herein are material and are of the essence of this contract.

The contract provided that A&S would be liable for liquidated damages as a result of any delay until the project was substantially complete. Metro reserved the right to refuse payment if Metro was of the opinion that A&S's rate of progress was such that substantial completion of the project would be delayed.

The contract provided that if A&S was delayed in performing a critical task because of some act or omission of Metro or someone acting on Metro's behalf, including an authorized change order, then A&S could submit a written request for an extension of time with Metro or its architect or engineer, setting forth in detail all known facts and circumstances supporting the claim.

Finally, the contract provided for two methods of termination by Metro. The first allowed Metro to terminate the contract for convenience. If the contract were terminated in this fashion, A&S was to be compensated for labor, material, equipment, and services accepted under the contract, and certain costs that had been incurred in preparing to perform and in performing the terminated portion of the work.

The second method of termination by Metro was "for cause." If A&S failed to perform any portion of its work in a timely manner; failed to supply adequate labor, supervisory personnel, or proper equipment or materials; or violated any material provision of the contract, then Metro had the right to terminate A&S's performance, assume possession of the site, and complete the work. If the costs and expenses of completing the work exceeded the contract price, then A&S was liable for the excess costs. If such costs and expenses were less than the unpaid amount of the contract price, then A&S was to be paid the unused amount.

Several provisions of the contract should have been cause for concern to A&S. The owner's right to not pay A&S if the owner was of the opinion that A&S's rate of progress was such that substantial completion of the project would be delayed should have been a cause for concern because nonpayment would probably result in delayed completion. The termination clauses also pose high risks.

Contractor Execution

A&S was given notice to proceed under the contract on April 1, 2003. From the beginning, there were significant delays on the project. Excavation at the site was initially delayed because a house located on the site had not been moved. A&S was not allowed to demolish the house because it had been given to a private individual and Metro was waiting for that individual to make arrangements to have it moved. The house was finally moved on April 23, 2003. A&S requested a 15-day extension of time because of the delay. The project architect responded to this request stating, "The additional time of 15 days is approved. It has been stated by the Owner that the original date of substantial completion should remain intact until the end of the job, at which time, a change order will be issued if the additional time is needed to complete the work."

Once the site preparation began, it was immediately discovered that poor soil conditions not anticipated by the contract documents would require corrective measures. It was not determined what would be done to remedy this condition until August 26, 2003 (approximately four months later), when the parties agreed to lower the elevation of Oliver Middle School and the surrounding area to provide fill material for the Shayne Elementary School building. A&S requested an additional 110-day extension of time, including "71 days for soil, 15 days for house removal delays, 3 days for providing additional drainage at Shayne Elementary, and 21 days for additional undercut of soil on the site . . ." A&S signed Change Order

No. 1 (CO1) on September 10, 2003. CO1 provided for an extension of 50 days, and the date of substantial completion was set as July 15, 2004.

Because of the delay in the commencement of the project, the masonry work, which represented a significant portion of the work to be performed, was pushed into the winter months rather than the summer and fall months as originally contemplated. By mid-January 2004, it was apparent that the project would not be completed on time. About that time, Metro's architect was said to have admitted that the July 15, 2004, completion date was not based on what was practical, but only on the desire of Metro to have the schools ready for the beginning of the 2004–2005 school year. After that time, however, Metro continued to revise the scope of the work but steadfastly refused to alter the substantial completion date.

On February 23, 2004, the parties executed Change Order No. 2 (CO2). This change order required A&S to remove and replace an additional 15,436 cubic yards of unsuitable soil at the Shayne Elementary site. A&S requested an additional 21 days to perform this work, but Metro refused to grant any time extension. Before executing CO2, A&S notified Metro in writing that A&S was unable to sign the change order because of the denial of the requested extension of time. A&S explained that in an effort to keep the project moving forward, A&S would sign the change order with the understanding that A&S reserved the right to recapture the days lost during performance of the work. A&S ultimately signed CO2.

From mid-February 2004 through the end of July 2004, Metro continued to make changes in the scope of the work. There were 13 letters from A&S to Metro requesting a total of 235 days additional time. No additional contract time was given by Metro. Three of the letters are worthy of note.

A letter dated February 17, 2004, quoted the cost to "make kitchen hood and mechanical equipment room piping and cooling tower piping changes" and requested 14 days additional contract time to perform the changes. The project manager (PM) for Metro wrote "Proceed" on the letter and signed it with the date June 9, 2004.

By letter dated June 15, 2004, A&S quoted the cost of providing ceiling diffusers at Shayne Elementary School and requested an additional 40 days of contract time after authorization for their installation. The PM returned this letter with the notation, "Proceed with installation. Time extension to be determined." This notation was dated June 28, 2004.

In the third letter, dated June 3, 2004, A&S quoted the cost of providing an outside air supply for three heat pumps in Shayne Elementary School and noted that the "mechanical subcontractor has requested 18 calendar days of contract time to perform the necessary work required to supply outside air to the heat pumps." On the same day, the architect instructed A&S to advise the mechanical sub to "[p]roceed with the work relative to adding a fresh air duct to HP-A1.2, and 3."

Despite Metro's PM and architect having seemingly approved time extensions, Metro continued to assert that the substantial completion date was July 15, 2004.

After that date, Metro allowed A&S to continue working on the project and continued paying for work performed. There is evidence that Metro revised the scope of the project five times during the month of July 2004, with three of the revisions coming after July 15, 2004. A&S and Metro agreed to alter the construction schedule to allow certain portions of each school to be constructed to accommodate students on the opening day of school. On July 27, 2004, A&S's mechanical subcontractor, by letter, quoted a price to Metro for acceleration at the Oliver School to complete its work there by August 8. The Metro PM accepted this proposal on July 27, 2004, and noted on the letter that the mechanical subcontractor's work at Shayne Elementary should be completed by August 2, 2004. On July 29, 2004, Metro agreed to a proposal by another subcontractor to accelerate fire alarm and monitoring installation and to complete their work by August 3, 2004. Also on July 29, 2004, Metro accepted a proposal from the plumbing and HVAC subcontractor to accelerate its work and complete it by August 8, 2004, at Oliver, and by August 2, 2004, at Shayne.

On July 29, 2004, Metro delivered to A&S a termination letter stating that, as of Friday, July 30, 2004, A&S would be relieved of their services because they had failed to meet the July 15, 2004, substantial completion date. A&S removed itself from the site and did no further work on the project. Metro completed construction of the schools by hiring a construction superintendent and primarily using A&S's subcontractors. A&S's PM stated that he was told by the director of plant planning and construction for Metro that A&S was terminated to recoup the additional costs of having A&S's subcontractors accelerate their work.

Many of the risks described in the above example were foreseeable. The written contract posed certain risks that could have been foreseen. The owner's resistance to a time extension was foreseeable. A careful examination of the geotechnical report might have suggested the presence of unsuitable soil. Furthermore, an inquiry of other contractors about their past experiences with Metro might have revealed that this owner or designer was not desirable to work with or was untimely in making decisions. Despite the foreseeable risks, A&S nevertheless chose to submit a bid. Is it likely that A&S made a profit on this project? The reputation of the company may have been permanently damaged. There may be a risk of insolvency.

3.7 Preliminary Planning Considerations

Develop a Preliminary (Leisure Time) Execution Plan

A preliminary execution plan (schedule) should be developed at the outset. This plan should be developed at a macro level. The format may be best done as a bar chart. This schedule is used to determine if conventional means and methods and sequences will allow the contractor to complete the project on time (if nothing goes wrong and if there are no surprises).

This preliminary schedule usually has several sequential operations and should reflect how a contractor would approach the job if given a generous amount of time to perform the work. This schedule establishes whether the plan needs to be cost oriented or time oriented. A cost-oriented schedule relies on conventional methods and often applies sequential scheduling practices. A time-oriented schedule is driven by unconventional methods and relies on concurrent scheduling practices. The plan for the State College Municipal Building (SCMB) is shown in Fig. 3-5. Because Fig. 3-5 relies on conventional methods and sequential scheduling practices, it is a cost-oriented schedule.

Fig. 3-6 illustrates a cost-oriented schedule versus a time-oriented one. It relies on the Outreach Building project shown in Fig. 3-7. Fig. 3-6 shows only two activities: exterior insulation and masonry. For a cost-oriented schedule, these two activities are done sequentially, that is, the insulation activity is completed before the masonry begins. Presumably, this is the least expensive way for masonry to proceed. It requires sufficient scaffolding to accommodate one masonry crew. In a time-oriented schedule, the masonry proceeds in two areas simultaneously. This method requires two masonry crews and more scaffolding, and it involves a faster consumption rate of brick.

A schedule is cost oriented if there is ample time for conventional means and methods to be used. However, a time-oriented schedule is usually preferred even if the time allowed by the contract is adequate. A contractor should try to complete a project as soon as possible because a longer schedule is exposed to more risk. Also, time is money.

Fortunately, there are a limited number of ways to build most projects and to develop a macro schedule. As more detail is developed in subsequent steps, the preliminary plan or schedule and general strategy can be revised. Table 3-1 presents some fundamental principles for developing and revising a preliminary schedule. The preliminary schedule is called a leisure time schedule. Fig. 3-5 is a leisure time schedule for SCMB. It is a 36-week schedule, which exceeds the allotted contract time by a factor of about two. Thus, the leisure time schedule is likely unacceptable.

Principles Applied in Developing the Preliminary (Leisure Time) Schedule— The preliminary execution plan should be done at a macro level (Principle 1.1) and should be based on the application of conventional means, methods, and equipment (Principle 1.2). Conventional means, methods, and equipment are probably economical and ones with which the contractor is most familiar. This schedule is developed early, when little schedule detail is available.

The preliminary schedule should make effective use of constructed areas (Principle 1.3), such as ground and elevated slabs, parking areas, and basements. Using constructed floors of a facility for material storage is not preferred; it can lead to the double-handling of materials and limits schedule flexibility. Materials and equipment will need to be removed from these areas before other construction activities planned for these areas are indicated.

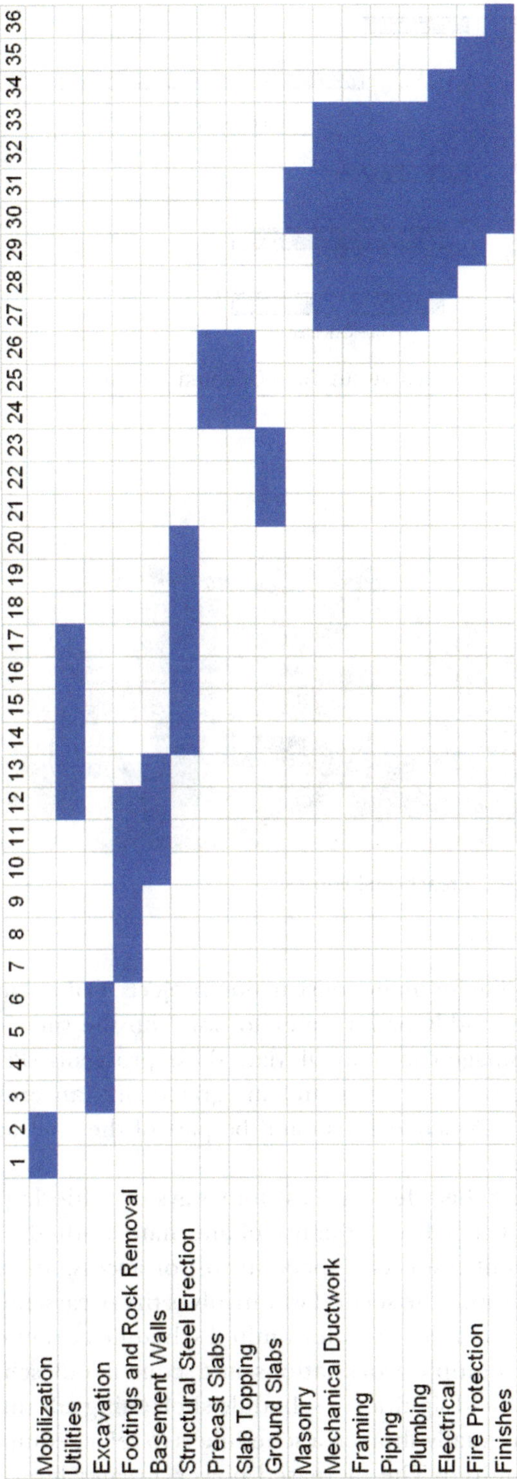

Fig. 3-5. Preliminary Execution Plan (Leisure Time Schedule), SCMB

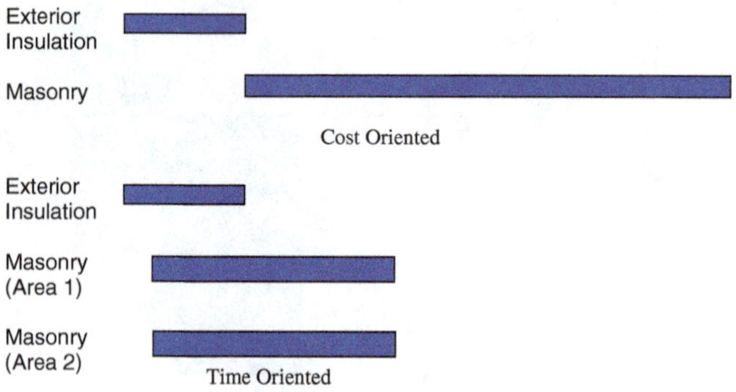

Fig. 3-6. Cost-Oriented versus Time-Oriented Schedule

Fig. 3-7. The Outreach Building

Rain is a fact of life for many construction projects. Contractors can ill afford for the site to become muddy or for water to pond on the site. Installing and using permanent site drainage can help alleviate these problems (Principle 1.4). These facilities should be installed early and integrated into an erosion and sediment control (E&S) plan. Principle 1.4 should be part of the contractor's conventional scheduling practices.

Shortening the Schedule—At least four ways of achieving shorter schedules may be investigated. The first is the use of alternate methods. Different methods may require different resources, space needs, or access, and alternate methods may be more expensive. The second way involves the preassembly of components. These first two alternatives may offer limited schedule advantages. The third way involves applying multiple workstations and concurrent scheduling practices. This approach can yield significant schedule shortening advantages (see Fig. 3-6). A fourth way to potentially shorten a schedule is to select equipment that is larger than ordinarily required. The increased power, greater leverage, and increased

Table 3-1. Fundamental Principles of Developing and Revising a Preliminary Execution Plan

No.	Principle
	Developing and Revising Preliminary Schedules
1.1	Plans and schedules should be done at a macro level.
1.2	Preliminary plans and schedules should be based on the use of conventional means, methods, and equipment.
1.3	Make effective use of constructed and footprint areas.
1.4	To allow more room and to mitigate rain runoff, install utilities and permanent site drainage first.
	Shortening Preliminary Schedules
2.1	Minimize storage area needs by off-loading directly from the delivery truck.
2.2	Minimize storage area needs, enhance the schedule, and improve labor performance by preloading materials onto each floor or area.
2.3	Make use of service elevators to transport lightweight or bulky items like duct and drywall to each floor or area.
2.4	Develop multiple site layout plans.
2.5	Sequence activities so as to avoid making noncritical activities critical.
2.6	Execute activities concurrently, rather than sequentially.
2.7	Seek opportunities to offer input into design.
2.8	Minimize the number of times a crane has to be moved.
2.9	Involve subcontractors in developing operational plans.
2.10	Unless necessary, do not plan the same activity to occur on the same day of subsequent weeks, for example, we plan to place concrete every Thursday.
	Developing Operational Plans
3.1	Partition the site into work area or zones.
3.2	Use multiple workstations to accelerate the work.
3.3	Vary crew sizes and teams to speed up or slow down the pace of the work and reduce crew idle time.
3.4	Integrate the work of multiple contractors.
3.5	Apply time lags or buffers.
3.6	Schedule the work to accommodate continuity of work of key resources.
3.7	Begin work in a location where the work is the hardest because this work takes the longest time.
3.8	Begin the work with the critical path work.

reach of oversized equipment may offer important schedule advantages, depending on the activity.

On SCMB, a quick and efficient method of removing rock from the footing trench needed to be found. Blasting could have been considered for the revised preliminary schedule. The schedule improvement and increased labor efficiency might have more than offset any increase in cost. Multiple workstations and concurrent scheduling should have been adopted.

Decide on Key Macro Strategies

Decisions must be made on two categories of macro strategies: contractual and general. At this point, it is known whether the revised schedule will need to be cost oriented or time oriented and if the contractor can finish the work on time.

Contracting Strategy—Some activities are specialized and can best be done by specialty contractors or subcontractors. The work may involve special equipment or labor and management skills. The work may require close vendor coordination or scheduling that is best suited to a specialty contractor. Examples include curtain wall erection, steel erection, precast erection, blasting, and much of the interior service work, such as mechanical, piping, and ductwork. Activities requiring special skills or knowledge are often best left to subcontractors.

General Strategy—In developing a general strategy, a contractor must decide where the work will begin and how the work will proceed.

There should be a reason for starting the work in a particular place. The footprint area should then be partitioned into work areas or zones to facilitate this strategy. Avoid partitioning into too many areas. The direction of the work progression, say clockwise or from one area to another, should be established.

It is probably beneficial to start work on the difficult-to-complete activities because these activities take the longest amount of time (Principle 3.7). If there is adequate time lag between the activity and follow-on activities, congestion and idle time will be reduced or eliminated. Be sure that this work is on the critical path (Principle 3.8). The work should also be on the facility being built, not peripheral structures like retaining walls, unless there is good reason to do so. Sometimes it may be necessary to start on peripheral structures to mitigate certain risks. Try to do work on the facility concurrently with mobilization, peripheral structures, drainage structures, and site utilities. Concurrent scheduling practices and partitioning of the worksite offer significant schedule benefits.

The direction the work flows is an important part of the general strategy. If work starts concurrently at opposite corners of a structure, it may be possible to avoid congestion, depending on daily production rates. The general strategy is unique to each project, but it should be simple and should accelerate the work.

Examples—Consider the construction of a three-story, 97,000-ft^2 office or classroom building known as the Outreach Building in State College, Pennsylvania (see Fig. 3-7). The project contains a 50-plus-ft curtain wall and about $15 million of electronic and satellite equipment. Both the curtain wall and the equipment installation were subcontracted. The building had to be enclosed and sealed before equipment installation could begin.

The footprint area was partitioned into two areas, right and left, with the work progression flowing from right to left. As a general strategy, the contractor could begin the foundation work (footings, piers, and grade beam) on the half where the curtain wall is located (the right half). This strategy expedited the structural steel erection for the curtain wall and masonry in that area and may have expedited the installation of the curtain wall. If successful, this strategy might have allowed the building enclosure to occur sooner, meaning that electronic equipment installation could have been expedited. This simple example illustrates how general strategies can be applicable to all sizes and types of projects, even small, simple ones. If everything goes as planned, the schedule may be shortened because the work began on the right half.

As another example, consider the SCMB. One excavation strategy might be as shown in Fig. 3-8. The following strategy recognizes that most of the rock in the footer trench is on the northwestern and western sides (point A to point B). There may be some rock from point B to point C. The earth excavation is followed by (1) rock removal and trench excavation, (2) footing construction, and (3) concrete wall construction. The footprint area (for the earth excavation) is divided into four parts. The earth excavation begins in Area I (at point A) and proceeds in a counterclockwise manner followed soon thereafter by the work in Area II. As the excavation approaches point C, a second rock removal and trench excavation crew can begin work at point B. The footer and wall formwork crews follow the rock removal and trench excavation crews, beginning at point A. An appropriate time lag is applied because footing construction is much faster than the rock removal and trench excavation activity. The trenching work from point B to point C is much faster than the trenching work from point A to point B because there is little or no rock. Footing construction should not "catch up with" the rock removal and trench excavation activity. This strategy allows all crews to proceed at an unimpeded pace, a practice that promotes efficiency. The practice of concurrent scheduling and multiple workstations (for rock removal and trench excavation) expedites the construction of footers and the basement wall, which are critical path activities. This simple strategy may shorten the overall construction schedule by a month or so.

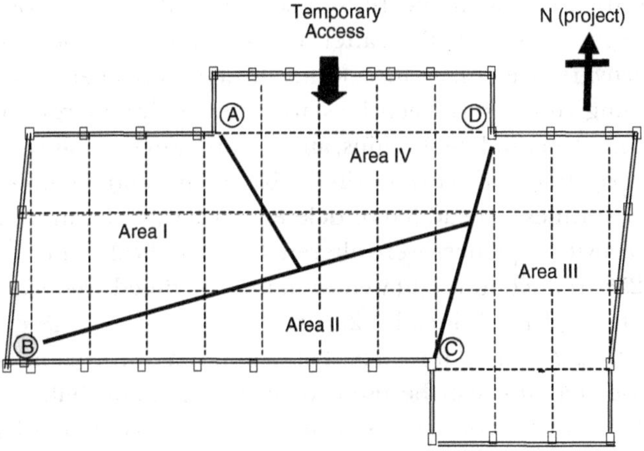

Fig. 3-8. Partitioning of SCMB (Excavation Only)

These two examples show that where the work is started is not a trivial matter. The key principles used are beginning where the work is hardest, partitioning, concurrent scheduling, and using multiple workstations. The benefits are that the time the work is exposed to inclement weather is reduced, and the potential for schedule shortening is significant.

Determine Major Means, Methods, and Equipment Needs

At this point in the planning process, one should select major means, methods, and equipment needs. It is important that the selection be compatible with the schedule orientation. To illustrate this point, a crane and bucket operation is often used for concrete placement. It is a relatively low-cost piece of equipment, although highly inefficient. If the schedule is time oriented, a pump truck may be a preferred alternative. On the SCMB project, removal of rock using hand tools (the practice that was actually applied) is consistent with a cost-oriented schedule. If utilities are done early (see Fig. 3-3) and the work begins at point A (see Fig. 3-8), all wall concrete placement can be discharged directly from the delivery truck. Selection of major equipment should be done at this time to ensure availability. In some instances, renting equipment may be advantageous.

Revise Preliminary Execution Plan

At this point in the planning process, the preliminary execution plan may need to be revised to conform to the time requirements of the contract or some desired project duration. Fig. 3-9 shows a revised preliminary schedule where activities are executed concurrently rather than sequentially (Principle 2.6). There has been a 25% reduction in the scheduled weeks, even though the activity durations are the same.

The main differences between the schedules in Fig. 3-5 (preliminary) and Fig. 3-9 (revised) are that in the revised schedule, the excavation can begin as mobilization begins or shortly thereafter, excavation–rock removal–basement wall construction activities are done concurrently, structural steel erection–precast floor planks–preloading (not shown) activities are integrated, underground utilities are completed before structural steel begins, and the order of cast-in-place (CIP) ground slabs–masonry partition wall activities (in the basement) and the precast slab activity is changed. No activities were added or deleted, no activity durations were changed, and only four activity sequences were altered, hardly radical changes. Yet, there has been about a 25% reduction (eight weeks) in the scheduled (reduced from 36 to 28 weeks) simply by applying Principles 2.5 and 2.6.

Principles 2.1, 2.2, and 2.3 form a part of a comprehensive material management plan. The service elevator can be used for lightweight and bulky items. Loading directly from the delivery truck and preloading are practices that will improve labor productivity and schedules and will minimize on-site storage requirements (see Chapter 6).

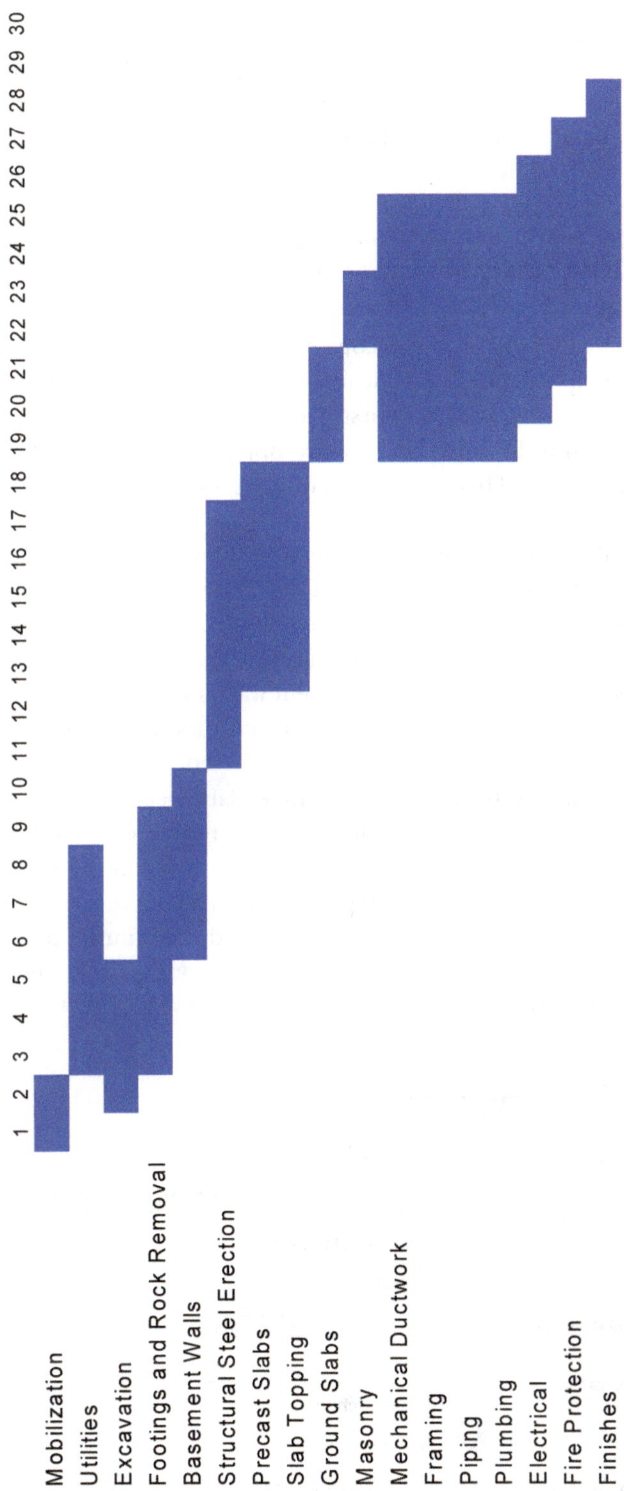

Fig. 3-9. Revised Preliminary Execution Plan, SCMB

Principle 2.4 states that contractors should develop multiple site layout plans as required (see Chapter 4). Plans should be simple and should require minimal change from one phase to the next. Site plans ensure the optimum use of the site.

Principles 2.5 and 2.6 relate to the sequencing of activities. The application of these principles can lead to the significant shortening of schedules. A fundamental principle is that noncritical activities should be delayed as long as practical. Activities like building retaining walls, grade slabs, and basement ground slabs are not usually critical activities, yet they are often sandwiched between critical activities. This normally makes them critical activities when they need not be critical.

On critical activities that require considerable time, consider working multiple workstations (Principle 3.2). Theoretically, the schedule can be shortened considerably. The schedule through the first half of many building projects is often observed as being mostly sequential. Many opportunities to work a concurrent schedule are overlooked. This principle is often ignored, despite obvious schedule advantages.

To illustrate Principle 2.5, consider the scheduling of three activities shown in Figs. 3-10 and 3-11. The three activities are the concrete basement wall construction, steel erection, and construction of the grade slab. Critical activities are shown in the dark color. Because there is only one concrete crew, the basement wall and grade slab cannot be done simultaneously. If the basement wall and grade slab are done back to back (to get all the major concrete work completed), then the grade slab is a critical activity and its duration lengthens the overall project schedule. This schedule assumes that steel erection cannot begin until the grade slab is complete. If, however, it is possible to delay the work on the grade slab and erect steel immediately after the basement wall is finished, then the grade slab is no longer on the critical path. This strategy requires that a scheme be developed for placing concrete for the ground slab.

Principle 2.8 states that cranes should be moved a minimum number of times. Whenever a crane is moved, the crew(s) using the crane have nothing to do. Idle time increases, and costs soar. Crane relocations may take two to three hours or

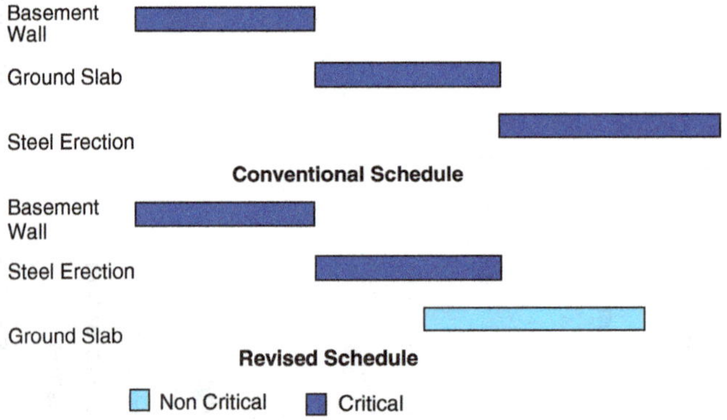

Fig. 3-10. Illustration of How Sequencing Affects the Critical Path

Fig. 3-11. Basement Wall under Construction, SCMB

more. Crane locations should be planned carefully. In some instances, crane relocations can be minimized by using a larger crane than needed.

Principle 2.9 is obvious but is often ignored. If subcontractors are expected to fully commit to a schedule, they must be involved in making the schedule.

A practice often adopted is to reserve a key activity to be done the same day each week. Such as, "we place concrete each Thursday." Such a practice is not always wise (Principle 2.10). Suppose that it takes four and a half working days to make the necessary preparations to pour concrete. What will happen the other half day? The crew will do busy work or cleanup. This half day will be wasted. A short interval production schedule (SIPS) is helpful in planning the appropriate cycle time.

3.8 Recapitulation

At this point in the planning process, it is worthwhile to summarize what has occurred. An affirmative decision to submit a bid has already been made. Most project risks are known. Planning began with the development of a leisure time schedule. By comparing the duration of this schedule to the contract requirements or some desired duration, it was determined whether the schedule is cost oriented or time oriented. Key strategies, means, methods, and equipment were selected that are compatible with the schedule orientation. A revised schedule that relies on doable means, methods, and equipment is produced that complies with the desired time requirements. Further shortening of this schedule may be accomplished by shortening activity durations, although an occasional resequencing may be justified. From this point forward, the main focus is to produce optimum work plans to make the work as efficient as possible.

3.9 Detailed Plans

Decide on Material Delivery Schemes

Material delivery schemes have much to do with how the site is used and the efficiency of site operations. The contractor may choose to deliver most or all of the materials to the site before beginning work. This scheme would require large amounts of storage space. Alternatively, the contractor may choose to erect certain materials directly from the delivery truck. While storage space will be lessened, delivery routes must be carefully planned. Preloading is a practice that can be highly efficient, but it must be carefully planned (see Chapter 6). Carrying materials up the stairs, one component at a time, is inefficient. Off-site staging areas or surge piles can sometimes be effectively used.

Develop E&S and Site Drainage Plans

In some locales, an erosion and sediment control (E&S) plan must be filed and approved by a local governmental entity or regulatory body. The E&S plan deals largely with runoff from the site, but there are more elements that constitute a complete site drainage plan. So the contractor should integrate other elements with the E&S plan (see Chapter 5) to produce a comprehensive site drainage plan. There are three elements of site drainage that should be addressed: (1) drainage of the facility footprint, (2) drainage of the remainder of the site, and (3) runoff from off-site areas into the site.

The application of Principle 1.4 contributes to site drainage and reduces the likelihood of water ponding on the site. The facility footprint is often ignored, but runoff that collects in footings can lead to significant disruption and rework. Shallow drainage trenches are probably adequate. These trenches should be installed before the need arises because rainfall does not always occur during normal working hours. A sump pump may be needed. A system of shallow trenches is probably sufficient to drain runoff from storage areas and other site areas. The E&S plan may require that the runoff be channeled into a retention basin before the water is discharged from the site.

Drainage from places such as off-site parking areas and steep slopes is often overlooked. These areas may funnel large amounts of water onto the site. A simple, shallow trench or ditch will probably solve the problem (Thomas and Ellis 2009).

Understand Environmental and Regulatory Issues

Environmental and regulatory considerations are an important part of being viewed by the public as a responsible contractor. Disregard for these requirements can easily damage the company's reputation and can lead to fines and lost profits. Additionally, there is a clause in all standard form contracts and in most nonstandard form contracts that requires the contractor to comply with all laws, ordinances, and regulations. Failure to comply constitutes a breach of contract. The project engineer and project manager need to be aware of a voluminous collection of federal, state, and local requirements. Ignorance is not an option.

A construction project requires various permits. Not all permits are acquired by the contractor, but compliance is still required. It is therefore incumbent on the contractor to know what permit commitments have been made by others.

Local ordinances may limit noise. You may be asked to do some things that are not in ordinances, such as dust control or adjusting the hours of work (e.g., not working before 7:00 a.m. in residential areas).

Statutes, Ordinances, and Regulations—It is not possible to list all the laws, ordinances, and regulations with which the contractor must comply. Federal, state, and local laws and regulations range from OSHA regulations, minimum wage rate requirements, and worker's compensation, to local ordinances, such as limits on noise levels. The contractor must be aware of the regulatory requirements. There is hardly a part of the contractor's site management practices that is not touched in some way by a regulation.

Permits—There is also a proliferation of permits that are required. Only four are mentioned herein. These are the erosion and sediment control (E&S) permit, wetlands permit, blasting permit, and building permit. Full compliance by the contractor is required.

E&S Permit

E&S permits may not be required in all locales, yet it is frowned upon to discharge sediment into a stream, especially an environmentally sensitive one. State pollution laws and regulations may apply.

Wetlands Permit

A wetlands permit is required from the U.S. Army Corps of Engineers before infringing on a wetland. Wetlands are not entirely defined by the amount of water that is present, but the definition also takes into consideration the type of vegetation and type of soil. Other criteria may be applied. Fig. 3-12 shows the economic consequences of noncompliance.

Blasting Permit

In most places, blasting operations need to be supervised by a licensed blaster. There are many other restrictions on blasting operations. The authors have observed numerous projects that were plagued by ineffective blasting operations, so a skilled blaster is a valuable resource who can save the contractor time and money.

Local firm fined for wetland violations

[redacted] has been fined $45,000 for wetlands encroachment at a natural gas drilling facility in Clearfield County.

In addition to the fine, the company has to cease all operations at the Cooper Township location and has 30 days to submit a remittance plan to the state Department of Environmental Protection.

[redacted] was fined after DEP inspectors found that two of the company's facilities — a natural gas well pad and a centralized impoundment pond — had been built in areas encroaching upon protected wetlands.

permit for its Alder Run natural gas well site, which had previously been approved by DEP.

In addition to the encroachment issue, the company also constructed the Alder Run centralized impoundment differently than what was approved in the permit, and that also affected a wetland.

[redacted], DEP spokesman, said he did not know whether the impoundment pond contained fresh water or frack water — a by-product of the hydraulic "fracking" process used to extract Marcellus Shale natural gas.

The company has 30 days to submit a mitigation plan and then six months to relocate its operations so they are no longer in violation, [redacted] said.

"I would assume what they would submit to us will be a reconstruction of the sites in accordance with their permit and (to) repair the part of the wetlands they encroached upon," he said.

Fig. 3-12. Consequences of Failure to Comply with Permit Requirements

Building Permit

Building permits are usually supplied by the contractor, but practices may vary. Other permits may be required. The contract needs to be consulted.

Develop Site Layout Plans

Now that drainage facilities are located and material delivery schemes are finalized, site layout plans can be prepared. Generally speaking, many contractors do a subpar job of site layout planning. In monitoring more than 200 projects over the past 35 years, the authors have yet to observe a project where there were comprehensive site plans. Additionally, research offers little insight into the mechanics of developing suitable site plans. A simple procedure is explained in Chapter 4. Through thoughtful site layout planning, the use of the site is optimized, and there is often more available space than is initially apparent (Principle 2.4).

Identify ETS Sequences

One does not have the resources or time to plan in intricate detail all the aspects of constructing a project. Fortunately, on most projects it is possible to identify one or more aspects of the project that are essential to success (ETS). In this step, ETS sequences, which may be a single activity or a string of activities, are identified, and the contractor's limited resources are marshaled to plan these sequences in detail.

ETS sequences differ from critical path activities in that in a CPM schedule, critical path activities cover the entire project, whereas ETS sequences may be only one activity or a limited number of loosely connected activities covering a limited portion of the project. Thus, an ETS sequence need not include a critical activity in the CPM vernacular. ETS sequences are unique to each project. The common denominator is that the sequence of activities must be done in an orderly or timely manner if the project is to be successfully completed.

Several ETS sequences are of utmost importance on most projects. For example, on building projects, it is always desirable to complete the foundation work quickly to minimize the risk of exposure to rain and mud. In colder climates, it is often desirable to enclose the building before winter weather arrives. In rainy climates, it is desirable to seal the building to prevent mold. There can be several ETS sequences on a single project, but probably no more than two to four.

On the SCMB project (see Fig. 3-2), two ETS sequences are identified. These are the following:

- Foundation: Excavation–rock removal–footings–concrete wall construction
- Superstructure: Structural steel erection–precast plank installation–preloading service and finish contractor materials.

The concrete foundation wall activities are selected because the foundation wall must be completed before the steel erection can begin, which is a critical activity, and because of the desire to limit weather-related risks. The rapid removal of rock is essential, and the selection of an appropriate method of rock removal is of paramount importance.

The steel erection–plank installation–preloading activities are also selected as an ETS sequence because the work of multiple contractors must be integrated. If steel is completely erected first (sequentially), then inefficient work methods will be needed to install the floor planks and to deliver the service and finish materials to each floor. These three activities may not be linked in a CPM schedule, yet how this work is integrated will have much to do with how productive and efficient the service and finish work will be done. If a few more activities are added, this ETS sequence can culminate in the building being enclosed and sealed to allow for temporary heat and to prevent water infiltration.

Calculate Time Windows

The detailed scheduling process includes a CPM analysis. A CPM schedule yields early-start and late-finish times for each activity in the revised execution plan. These times (dates) establish the boundaries or windows in which each activity can be scheduled.

CPM texts often refer to early-start and late-start schedules. The authors have never observed either schedule being used on a construction project. Both have significant disadvantages relative to time and cost. An early-start or late-start CPM schedule is likely not an optimum schedule. For an early-start schedule, there is no continuity of key resources, resource leveling may be needed, and the work of multiple contractors may not be satisfactorily integrated. There may be other deficiencies. A late-start schedule has little slack time, and most delays cannot be tolerated. For these reasons, the CPM schedule is the starting point, not the end point in the scheduling process.

At the end of this step, one should have a reasonable schedule that will allow timely completion and time windows in which noncritical and critical activities can be scheduled. Equipment and methods should be mostly finalized.

Develop Detailed Operational Plans

Most of the schedule reduction in the revised schedule has been achieved through specifying multiple workstations, applying concurrent scheduling practices, sequencing to minimize the number of critical activities, and selecting proper equipment and methods. Further schedule reductions can be achieved largely through the shortening of activity durations. The main goals of developing detailed operational plans can be many and varied. Some of the more common goals are to (1) accelerate the schedule, (2) ensure an efficient work plan and methods, (3) determine resource needs, (4) ensure continuity of work for key resources (reducing idle time), (5) determine time lags, (6) integrate the work of multiple contractors, (7) ensure safety and quality, and (8) ensure access.

Some principles for developing operational plans are given in Table 3-1. As a minimum, detailed operational plans should be developed for all ETS sequences.

Multiple tools can be used to develop operational plans. Some common tools are listed in Table 3-2 along with some of the major functions and usages. The most common tools are (1) bar charts, (2) CPM schedules, (3) short interval production schedules (SIPS), (4) linear schedules, (5) velocity or production rate charts, and (6) sequence plans. These tools (except for bar charts and CPM schedules) and Principles are discussed in the following.

Partitioning—The work area may be partitioned into zones or work areas (Principle 3.1). The partitioning may need revision as the details of the operational plans are developed. Multiple work areas facilitate multiple workstations, which facilitate concurrent work practices (Principles 2.6 and 3.2). Principles 2.6 and 3.2

Table 3-2. Tools Used in Developing Operational Plans

Tools	Function
Bar chart	Communicate the work plan.
CPM schedule	Determine project duration and calculate time windows.
SIPS	Accelerate the cycle time or schedule, integrate the work of multiple contractors, draw attention to efficient work methods, define needed resources and feasibility, and ensure continuity of work for key resources.
Linear scheduling	Ensure continuity of work for key resources, draw attention to appropriate crew sizes and avoid idle time, and avoid congestion.
Velocity charts	Determine time lags and ensure continuity of work for key resources.
Sequence plans	Combine all planning knowledge into an efficient work plan.

cannot be easily applied without the application of Principle 3.1. Collectively, partitioning and using multiple workstations can accelerate the work by allowing work to progress on a single activity in multiple locations. Partitioning also supports the development of site plans, facilitates the creation of linear and SIP schedules, and aids in determining minimum time lags.

Fig. 3-8 shows the partitioning scheme for the earth excavation only on the SCMB. It is partitioned into four areas. Fig. 3-13 shows the partitioning plan for rock removal and trench excavation, footings, and concrete walls. A review of the two portioning plans shows that the excavation, when done as prescribed, allows the work on rock removal and trench excavation to begin early (at point A). Excavation is always performed before other operations. The temporary access is the last area to be excavated. Two crews are used for the rock removal and trench excavation activity, and one crew is used for footing construction and wall construction. Finally, structural steel erection can begin on an expedited schedule. Figs. 3-8 and 3-13 are compatible with one another.

Short Interval Production Schedules (SIPS)— The short interval production schedule (SIPS) is a detailed plan that shows what work will be performed and when. It covers a short period of time (say, one to three weeks) and is often applied to cyclical work. It is not a look-ahead schedule because there is no element of hope in the SIPS. Once finalized, it becomes a commitment for all parties involved. A contractor is expected to add labor resources or work overtime to meet the production goals. A SIPS should be developed for all ETS sequences, and subcontractors need to be intricately involved in its development. The SIPS is useful for establishing appropriate crew sizes and teams to efficiently meet production commitments (Principle 3.3) and for integrating the work of multiple contractors (Principle 3.4). Principles 3.3 and 3.4 ensure that adequate resources are applied and that a realistic target time value is established. The SIPS may need to be modified as actual time values become known.

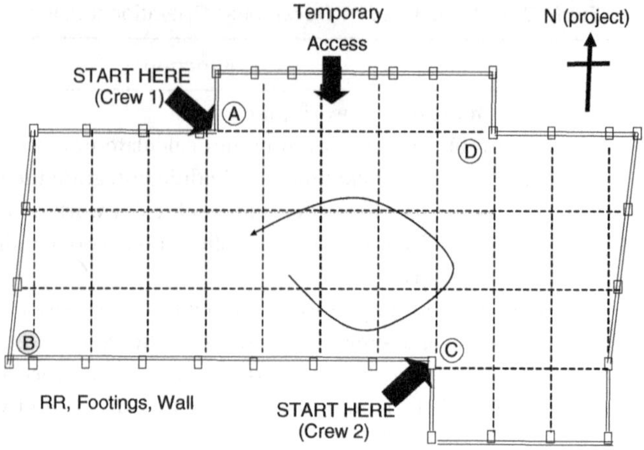

Fig. 3-13. Partitioning of SCMB (Rock Removal, Footings, and Concrete Basement Wall)

Example 1

An example of a SIPS for an apartment building is given in Fig. 3-14. On this project, it was determined that one ETS sequence was the completion of the roof by the end of November. To achieve this goal, it was necessary to complete one floor of the superstructure per week. The superstructure consisted of reinforced masonry walls (both interior and exterior) and precast concrete floor planks. Thus, the ETS was the superstructure: interior and exterior reinforced masonry walls–wall grouting–installation of precast floor planks–plank grouting–curing–preloading of masonry units for the next floor. The SIPS showed that this plan was barely feasible. The resources needed were (1) about 35 masons and helpers working at seven workstations; (2) a mortar and grout mix that cures to the required strength in about 18 hours during colder weather; and (3) regimented deliveries of block, mortar, grout, and floor planks.

On the basis of the SIPS, it was decided that the goal was feasible, although tight. Notice that the SIPS makes it possible to assess feasibility and the requirements for the plan to be doable. A word of caution is warranted. If the time for this work is made too liberal, it can cost the contractor money; flexibility to modify the schedule is needed. For instance, suppose one task is scheduled for three days in the SIPS schedule, but it can actually be completed in two and a half days. Unless the schedule is adjusted, the workers may stretch the work on that task to the full three days or spend the last half day in preparation for the next day or in cleanup (see Chapter 7). It is probably a mistake to schedule certain work to occur at the same time each week without the benefit of experience to ensure that the time allotted is suitable (Principle 2.10).

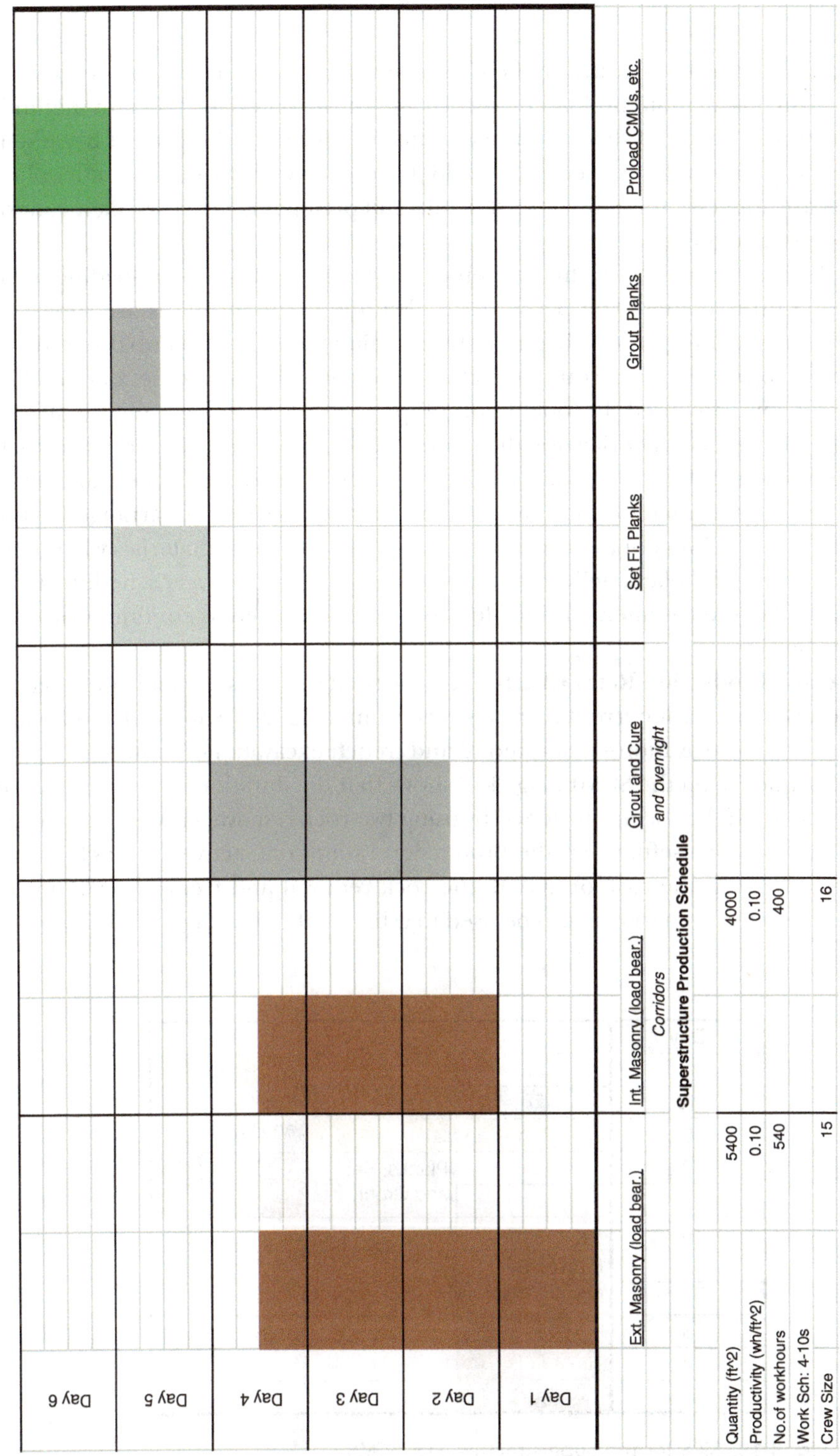

Fig. 3-14. SIPS for the Superstructure ETS, Bryce Jordan Tower

Example 2

This example is for the SCMB, and the ETS sequence is the structural steel erection–precast plank installation–preloading service and finish materials. Efficient methods and timely completion of this work are essential to success of the project because if the steel is completely erected (as would be the case if sequential scheduling practices are applied), then the precast planks and preloading activities likely cannot be done using a crane.

Fig. 3-15 shows a SIPS for the steel erection–plank installation–preloading to be done using three cranes. The partitioning plan and crane locations are shown in Fig. 3-16. The planning process begins by partitioning the work area into three zones. The plan is to proceed in a counterclockwise direction. In each zone, one level of steel is erected, followed by one floor of precast floor planks, and finally by preloading of service materials for that zone. Steel can be easily delivered to each crane location, and if the cranes are sized and located properly, only one crane setup for each crane is needed. Service and finish subcontractors need to arrange for the early and timely delivery of their materials. If they are late, their materials will not be preloaded. Steel and plank deliveries must also be on a strict delivery schedule. Duct will be stored in the basement (Principle 1.3). Notice that there is continuity of work for each activity.

Velocity (Production Rate) Charts—A velocity chart shows the daily production rate of one or more concurrent activities versus time. Fig. 3-17 shows a hypothetical example for the excavation–rock removal and trench excavation–footings–concrete wall ETS sequence on the SCMB. Fig. 3-17 shows that the duration of the foundation ETS for the SCMB has been shortened by using two rock removal crews. The velocity chart is particularly useful when the production rate of one activity is much slower than the others; in the case of SCMB, the rock removal and trench excavation is slowest. The velocity chart can also be used to estimate the time lag or buffer for the

Day	I.	II.	III.
11			Preload
10	Distribute Materials		Planks, Set and Grout
9	Preload Materials		
8		Planks, Set and Grout	Structural Steel Erection
7			
6		Align	
5	Planks, Set and Grout	Structural Steel Erection	
4			
3	Align		
2	Structural Steel Erection		
1			

Fig. 3-15. SIPS for the Superstructure ETS, SCMB

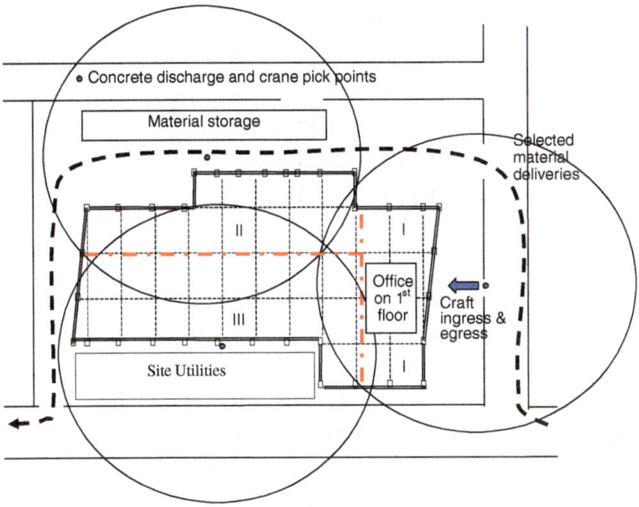

Fig. 3-16. Partitioning Plan, Superstructure Phase, SCMB

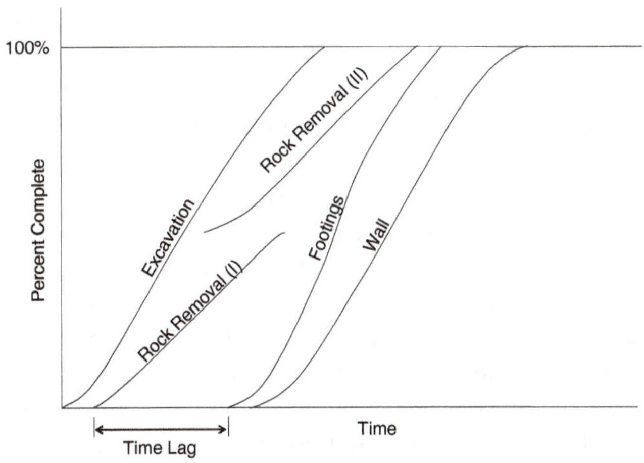

Fig. 3-17. Hypothetical Velocity Chart for Foundation and Basement ETS, SCMB

footing construction activity (Principle 3.5). Without time lags or buffers, all the ensuing work (footings and wall formwork) will be highly inefficient because of congestion. In this instance, the velocity chart can be used to evaluate various alternatives for accelerating the schedule, such as multiple workstations, multiple crews for rock removal and trench excavation, or faster methods. With partitioning and multiple work teams, time lags can be optimized or reduced to a minimum.

Linear Schedules—Linear schedules show the relationship between time and location. These graphical schedules are usually associated with linear projects, e.g., highways or pipelines, but they also have limited application on nonlinear projects in ensuring continuity of work by key resources, such as a concrete crew or crane (Principle 3.6). Ideally, to avoid inefficient downtime, a crew should proceed from one activity to the next without delay. Linear schedules allow one to visualize how to efficiently plan for a resource to be used to maintain continuity. The application of linear scheduling during the planning process minimizes crew idle time by ensuring that work is always available for the key resource. Harmelink and Rowings show how to identify controlling activities with linear schedules (1998).

To illustrate the linear schedule, the foundation ETS on the SCMB is used. Fig. 3-8 shows that the excavation work area was partitioned into four work areas (Principle 3.1). Note that if the work area is not partitioned, the development of a linear schedule for this ETS will be challenging. Two rock removal and trench excavation crews are to be used. Fig. 3-18 shows the linear schedule and a bar chart. Following some minor adjustments, it is possible to have continuity of work for the excavation and for the footing and wall formwork crews. Only one crew for each of these activities is required. The same cannot be said for the rock removal and trench excavation crew. A single rock removal and trench excavation crew will extend the schedule, and so two crews are planned (Principle 3.2). The crew sizes can be adjusted to ensure that the daily production goals are met (Principle 3.3). Time lags are also visible in Fig. 3-18. For example, the time lag in Area 1 between the completion of rock removal and the start of the footings is about 10 days (Principle 3.5). Without time lags, the work will become congested as fast activities catch up with slower ones.

Sequence Plans—Sequence plans as used herein show when subcontractors plan to work at any point in time. Sequence plans are usually thought of as pertaining to the service trades (e.g., mechanical, electrical, or drywall), but they may relate to any situation where the work of multiple trades or subcontractors needs to be coordinated. Sequence plans can be conveyed as a bar chart, SIPS, linear schedule, coordination drawing, or any other convenient format.

A hypothetical sequence plan was developed for the second floor of the SCMB. It is shown in Fig. 3-19. The first step in developing a sequence plan is to partition the work area. Fig. 3-20 shows how the second floor of the SCMB was partitioned. The corridor and other areas were partitioned into five unique areas. It may be advantageous for the scope of work in each area to be roughly equal, but this is not necessarily a requirement. The partitioning scheme shown in Fig. 3-20 is but one of many possibilities. For the sake of simplicity, only the following activities are shown:

Area I (corridor)

- Feeder duct,
- Feeder sprinkler pipe, and
- Feeder conduit.

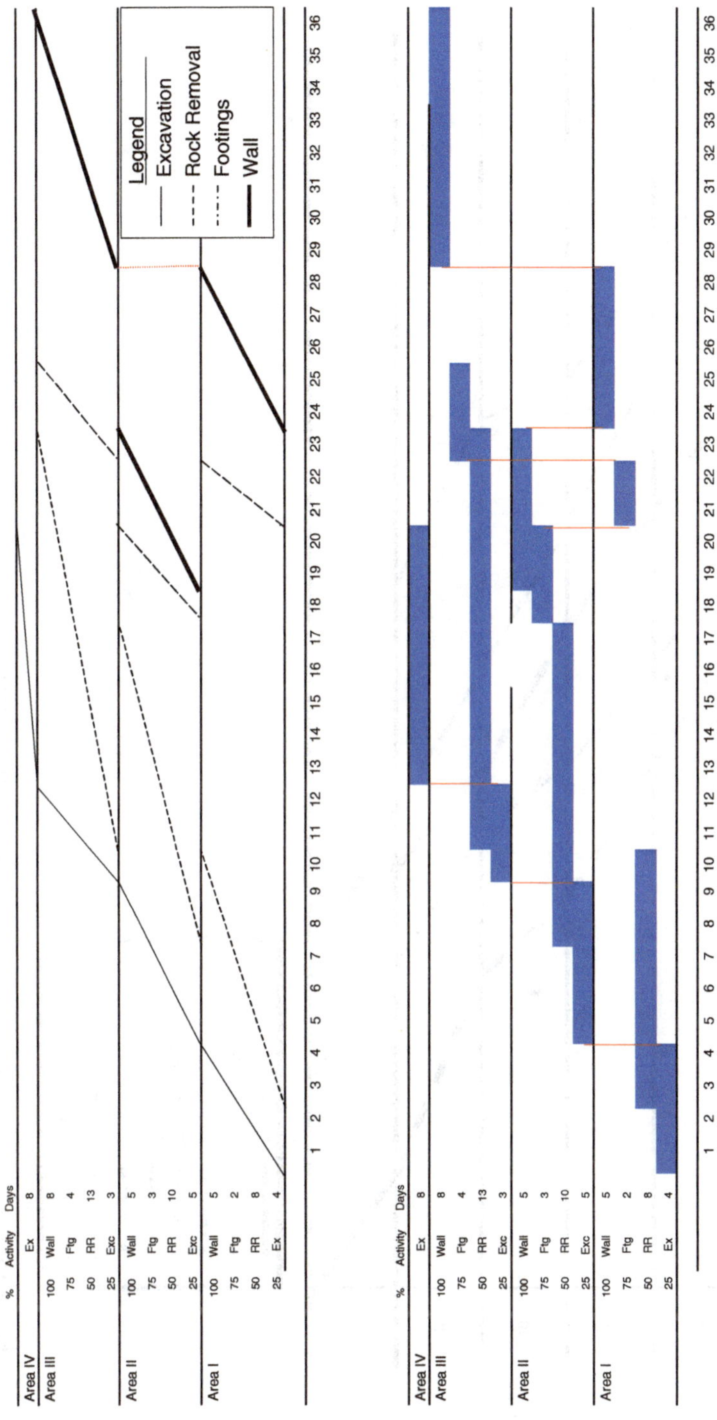

Fig. 3-18. Linear Schedule and Bar Chart for Foundation ETS, SCMB

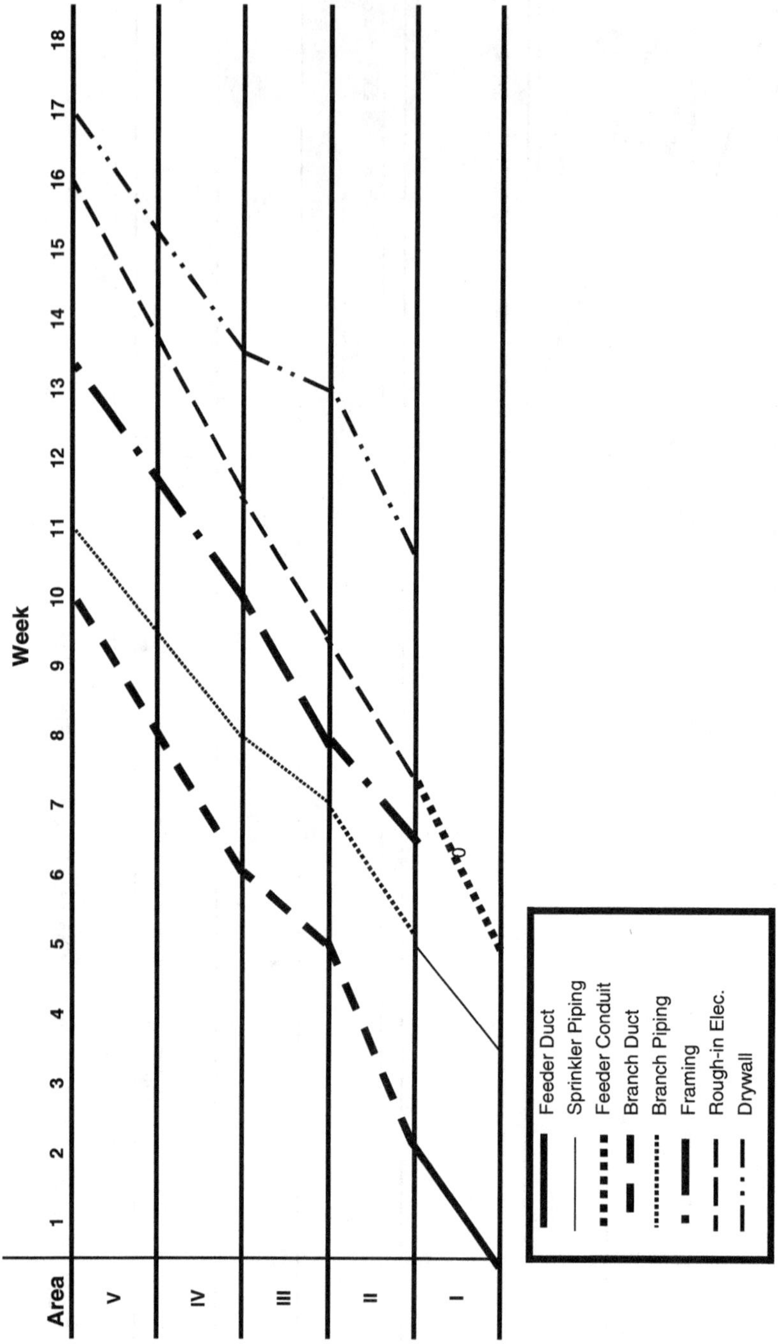

Fig. 3-19. Sequence Plan for the Second Floor Service Work, SCMB

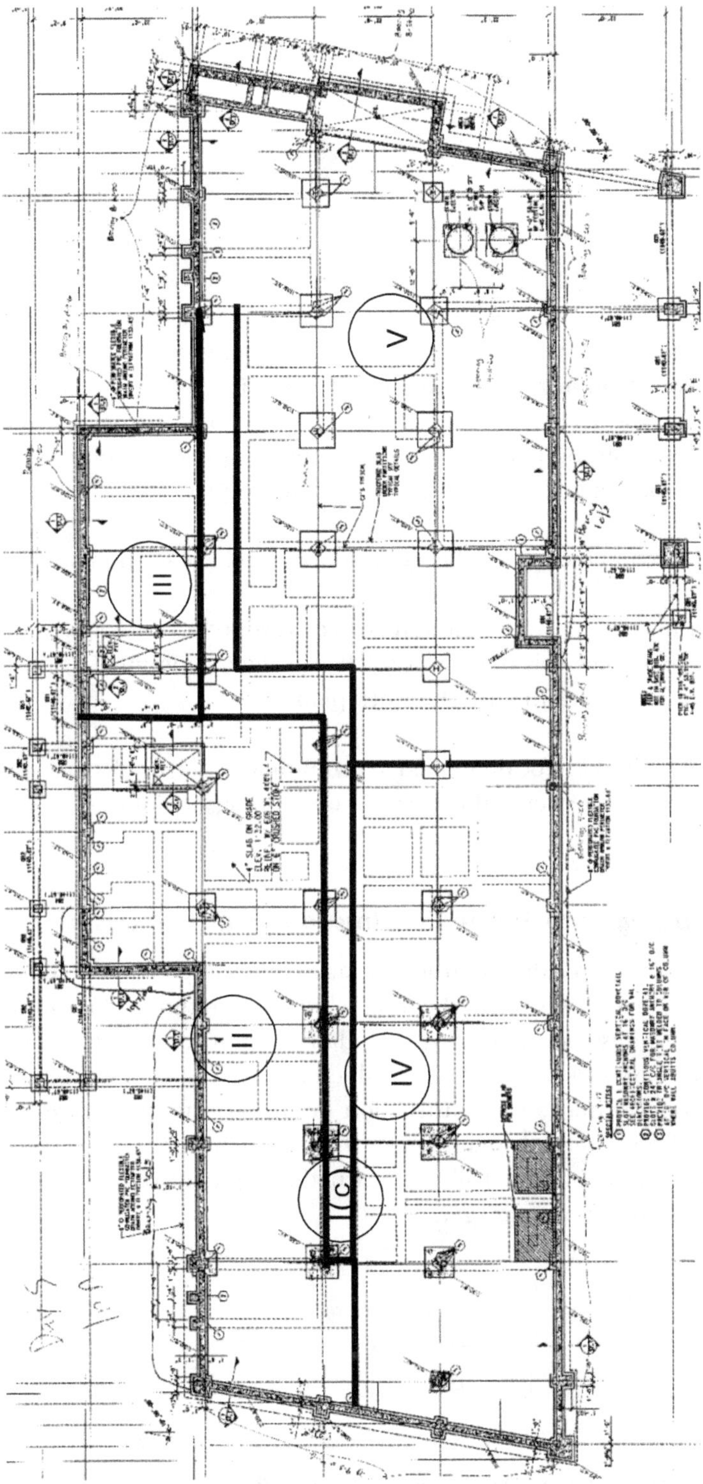

Fig. 3-20. Partitioning of the Second Floor, SCMB

Areas II–V

- Branch duct,
- Branch sprinkler piping,
- Framing,
- Rough-in electric, and
- Drywall.

There are several advantages of the sequence plan shown in Fig. 3-19. There is a one-week time lag between each trade subcontractor at all times, and no more than two subcontractors are working in any one area at any time.

Operational Planning Using Multiple Tools—The previous section described four scheduling tools. First, a CPM can be used to establish the schedule windows in which an activity can be performed without delaying completion. Then, the tools described can be applied to complete the planning process as follows:

- SIPS production control establishes resources and workstations and (cycle) time needed.
- Linear schedules ensure continuity of key resources.
- Velocity schedules (production rate charts) determine time lags.
- Sequence schedules finalize the work plan.

Lastly, a bar chart can be developed to communicate the results to field personnel. Fig. 3-21 shows the general process of developing operational plans.

Proactive Strategies to Ensure Construction Input into Design

The opportunity to provide construction input into design may be limited on projects that are competitively bid in the design-bid-build tradition. Substitutions can be requested of the designer. However, two other alternatives offer contractors opportunities to have input into the design: value engineering and alternative bidding.

Greater opportunities for construction input into the design are also possible with other delivery methods that engage the contractor early in the design phase. Two such delivery methods that allow construction input into design are design-build and using a construction manager.

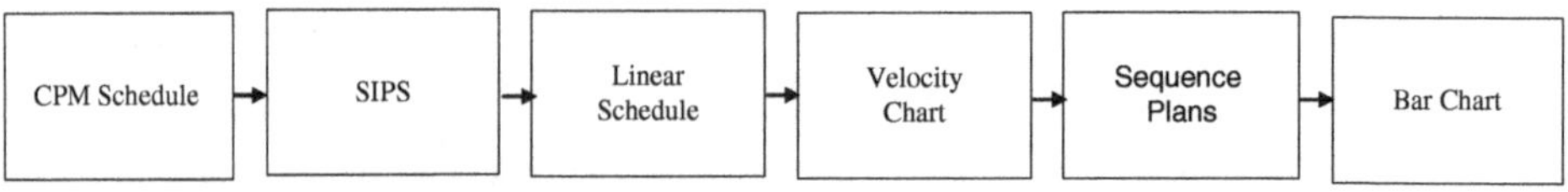

Fig. 3-21. Process of Developing Operational Plans

Substitutions

The contract sometimes allows substitutions if they are approved by the designer. These are usually thought of as equipment and material substitutions, but opportunities to substitute other aspects that can improve the schedule should not be overlooked.

Value Engineering

Where the contract allows the contractor to submit a value engineering (VE) proposal, the contractor may prepare an alternative design or proposal for the owner's consideration after the contract is signed. If accepted, the owner and contractor share in the cost savings.

The problem for the contractor with VE is twofold. First, there may be insufficient time for the contractor to investigate and/or design an alternative that leads to significant cost savings. Also, the owner may take an inordinate amount of time to review the proposal. Complex proposals that provide significant project benefits can be time-consuming to develop and review.

The second problem for the contractor with VE is the contractor's reluctance to pursue a complex VE proposal. If the VE proposal is rejected, the investigative and design expenses come off the contractor's bottom line (profit), and the contractor must absorb the cost.

Alternative Bidding

Another option is to include alternatives to the base bid. Bid proposals must always contain a base bid cost for exactly what the contract documents require; otherwise, the contractor's bid is not responsive to the solicitation and the bid may be rejected. However, the contractor should be able to say to the owner, "If you add, delete, or change *x*, then I will increase or decrease my bid by *y* amount." If an alternative is accepted, any design changes are now the obligation of the owner. The contractor has invested little of his or her own financial resources to propose the change, and any time delays are the responsibility of the owner.

Design-Build

Placing design and construction responsibility with a design-build contractor provides an ideal opportunity for construction input into the design. Teaming arrangements may vary from a single design-build firm, to a partnership between a designer and a contractor, to a joint venture. However, in all cases the construction team is engaged with the design team early in the design development process.

Using a Construction Manager

The construction manager approach to project delivery also makes construction expertise available during the design process. The construction manager's primary responsibility during design development is to provide construction input. This construction input is particularly important to the construction manager at risk, who will assume the role of a prime contractor and construct the project.

Finalize Execution Plan

It is important that the operational plans be compatible with the final execution plan (schedule). Because the operational plans are more detailed, the revised preliminary execution plan should be updated to reflect more realistic activity durations and sequences. The updated plan becomes the final execution plan. This plan should show that the project can be completed within the time allowed by the contract. Fig. 3-22 is a final execution plan. This plan shows additional shortening of the schedule compared with the revised preliminary schedule.

Further reduction to the overall schedule is still possible. Thus, the operational plan phase can result in the shortening of the durations of the foundation ETS and the interior service installations (e.g., mechanical, electrical, and piping).

Communicate and Enforce the Plans

How plans are communicated to others is of some importance. The means should be thorough but simple and clear. No doubt should be left as to what is to be done and when. A simple bar chart like the one shown in Fig. 3-23 is generally sufficient.

Develop a Submittal and Shop Drawing Schedule

After the intent of a contract award has been communicated by the owner, a submittal and shop drawing schedule should be prepared. Many projects have been delayed because the submittals or shop drawings were approved late. The main emphasis herein is on shop drawings because they generally take longer to draw up than ordinary submittals. Some items requiring shop drawings are needed early, such as foundation reinforcement. Other items may take inordinate review time, such as structural steel or precast components. Still other items require long lead times; windows are a good example. The important thing is to identify those items where the schedule is tight and begin the process early. Subcontractors should be involved in this process. Dates the shop drawings are due to the contractor should be incorporated into the schedule.

Certain submittals should not be overlooked. Mockups of masonry walls and other submittals may go through several iterations before approval.

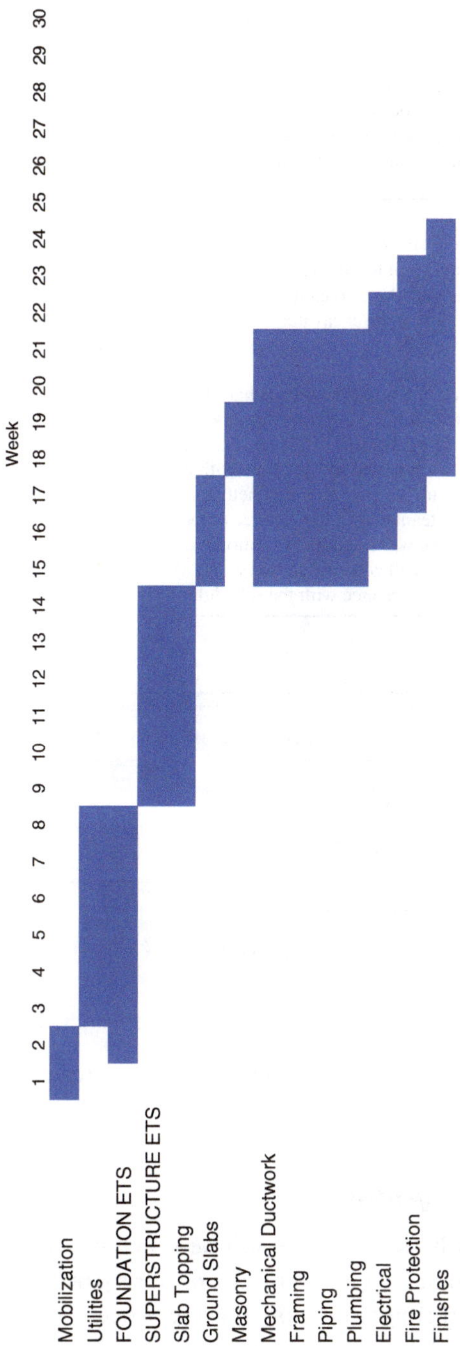

Fig. 3-22. Final Execution Plan, SCMB

Project: SCMB　　　　Date: Jul 6, 2005
ETS Sequence: FOUNDATION
ETS Activities: Excavation-Rock Removal-Footings-Concrete Wall

Description of Work:
Excavation—all earth to elev. 908.5
Rock Removal—removal of dense, hard limestone in trench to elev. 906.5; Approx. 350 ft. Trench is 2 ft 8 in. wide; rock is anticipated throughout entire perimeter
Footings—conventional footing concrete walls—conventional 9-in.-thick walls, to be placed in summer months

Methods and Resources:
Excavation—will be done via zones with one excavation crew supported by a front-end loader. Dump trucks will haul the spoil to the McHenry site (4 miles); no spoil is to be stockpiled on site. An earthen ramp is to be built to gain access to the basement area, but it is to be removed once excavation is complete. The excavated basement area will be used for rebar and formwork storage. Movement of materials is to be done via crane. Approx. excavation = 3,000 CY.
Rock Removal—the rock in the footing area is hard limestone (% recovery = 90%). The work will proceed at two workstations at opposite corners of the site. The removal of rock will be done with conventional jackhammers. This work is to begin one day after the corner of the footprint is exposed. Rock removal will be done with two teams of two laborers each.
Footings—this work will utilize conventional methods to construct built-in-place footing forms. There will be one team consisting of three carpenters.
Wall Formwork—this work will rely on conventional methods. One crew of six carpenters using 12' *x* 18' gang forms will accomplish this work.
All Work will be done in accordance with the schedule shown below.

Schedule:　　　　Duration: 36 working days

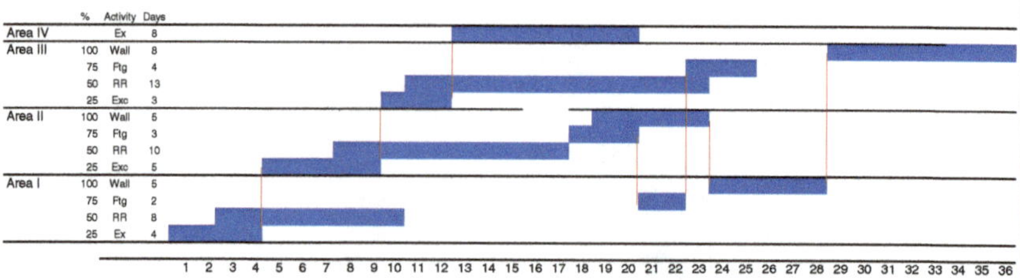

Fig. 3-23. Communicating Operational Plans

A submittal log should be developed to track the status of various submittals and shop drawings.

Plan for Change Management

Be wary of any owner who says, "There will be no changes on this job." There are always changes. It may signal an intent on the owner's part not to pay for changes by saying "fix the problem through the shop drawing process" or "that problem is one of means and methods." The contractor may be forced to engage in design work. Nevertheless, if the owner hints that there will be no changes, it may be enough of a red flag to suggest that the contractor should not submit a bid.

Most writings on change management speak to the importance of developing a change order log. A request for information (RFI) log is an equally important part of change management. These logs are used in contemporary writings primarily so that the contractor can prepare a claim. However, the logs serve an important function in the absence of a claim. They allow the contractor to inform the owner and designer which RFIs or change orders are negatively affecting the project schedule.

Contractors should be proactive relative to changes. The worst-case scenario is the need for a change to be identified by the crew in the field doing the work because this leads to much idle time and inefficiency. Therefore, contractors should strive to identify the need for changes well before the work is to begin. There should be a good process in place for incorporating changes into the work assignment before it is given to the supervisor. When a change proposal is submitted, the contractor should always reserve the right to claim impact costs because of labor inefficiencies and thus recover associated time delays.

3.10 Case Study—Millennium Science Complex

This case study illustrates that more attention to planning can lower labor costs and increase profits. The project is the Millennium Science Complex built on the Penn State campus in State College, Pennsylvania. The facility is a 275,000-ft^2 classroom and laboratory facility built in the 2008–2011 time frame at a cost of $215 million. A cm-agency delivery system was used.

Project Description

The building has a structural steel superstructure and is clad with a precast curtain wall and metal panels. The precast facade panels have a thin brick veneer, which gives the appearance of a brick building. The foundation is built on micropiles. The building is a four-story, L-shaped structure. An artist's rendering of the completed project is shown in Fig. 3-24.

Activity Description

The activity being reported on in this case study is the installation of the exterior precast panels. The work was done by a specialty contractor that was owned by the precast vendor. The panels were approximately 22 ft × 10 ft, although actual sizes varied. This meant that the panels were not interchangeable.

The panel installation crew was divided into two teams. The first team erected and aligned the panels and then secured the panels to the structure. This team consisted of 6–12 workers, although most of the time, there were 11–12 workers. The second team, which was much smaller, patched the lifting lug areas, installed

Fig. 3-24. Artist's Rendering of the Millennium Science Complex

insulation to the back of the panels, and caulked the joints. The scope of work of the second team was much smaller than that of the first team.

The fabrication yard for the vendor was about 150 miles from the project site. The superintendent ordered panels by 1:00 p.m. to be delivered the next morning at 6:00 a.m. Thus, the superintendent was gambling that certain panels would be erected the previous afternoon and that conditions would allow for the designated panels to be erected the next day. There was only one panel delivered per truck, and there were no panels stored on site. If the delivery of the panels was delayed the next morning, the erection team would be idle. If the team erected all the panels that were ordered that day in less than the time expected, and the superintendent was slow in sending the workers home, there was more idle time. In a few instances, the first team would help the second, even though the work needs of the second team's work were small. It did not seem that coordination between the contractor and vendor was as good as it could have been, and there were multiple opportunities for crew idle time and inefficient work. The first team may have been engaged in "busy work" some of the time.

During the latter part of the observation period, the work was not sequenced well. This problem was caused in part by steel bearing plates not being available or subpar vendor coordination. The welding of the bearing plates was done by another contractor. A subassembly should have been used, and the bearing plates should have been welded by the fabricator at the fabrication shop instead of in the field. To continue work, the crane had to be moved. Thus, there were more crane movements than planned, and this led to more crew idle time. Additionally, the crane movements did not seem to be planned for after work hours.

There was a lack of site planning to promote labor efficiency. Panel delivery trucks had to back into position. This problem can be observed in Fig. 3-25. Drive-through deliveries were not practiced consistently, leading to idle time. Fig. 3-26 shows further evidence of poor site planning. Because of the requirements of numerous materials and trash, the delivery and service trucks and crane did not have unrestricted access to the work face as the work progressed.

Fig. 3-25. Panel Delivery Practices on the Millennium Science Complex

Fig. 3-26. Panel Erection Showing Congested Work Area, Millennium Science Complex

Fundamental Principles

Principles 2.1, 2.6, and 3.8 were applied on this project. The following principles were not applied: Principles 2.4, 2.8, 3.1, 3.2, and 3.5. The following principles from

Table 3-1 were not observed or do not apply: 1.1, 1.2, 1.3, 1.4, 2.2, 2.3, 2.5, 2.7, 2.9, 2.10, 3.3, 3.4, 3.6, and 3.7.

Much of Chapter 7 is applicable to this case study. Case Study 3 in Chapter 7 should also be reviewed.

Contractor Performance

Fig. 3-27 shows the crew productivity for the precast panel erection. As can be seen, the performance for the first 13 workdays was rather consistent, and thereafter it was not so consistent. Lower numbers in Fig. 3-27 are better.

From the outset, the erection of panels was not done in a systematic fashion. There could have been two reasons. There could have been poor coordination with the vendor or the steel bearing plates may have been installed sporadically. Nevertheless, minimal crane movements were required the first 13 days. Beginning on workday 4, erection began on a cantilevered portion of the building. The weight of the panels led to deflections that required realignment of most of the previously installed panels. Productivity degraded through no fault of the contractor. Thereafter, panel erection became erratic, and numerous crane movements were required. The crew only worked six of the next 13 days. It appears that vendor coordination was somewhat of a problem.

One foreseeable risk is that subpar vendor coordination can lead to excessive crew idle time and inefficient use of labor. Site layout planning and effective operational plans that anticipated the vendor coordination risk would have minimized some economic losses resulting from this risk. It seems that the contractor would have benefited from a better plan.

It seems that the crew would have performed better if there had been better coordination (a planning issue) with the vendor. It also seems that there might have

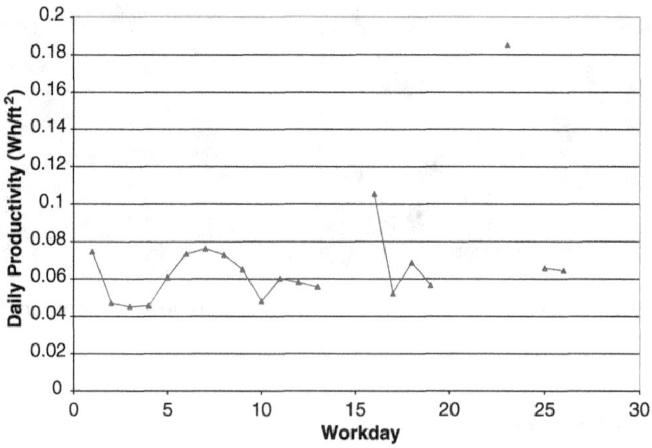

Fig. 3-27. Labor Productivity of Panel Erection Crew, Millennium Science Complex

been better performance had the panel erection crew waited longer (increasing the time lag) to start panel erection, allowing the bearing plate installation to proceed earlier and the vendor to produce a greater backlog of panels. The labor inefficiency is estimated to be 32%, and at $35/h (burdened), the contractor suffered a loss of $14,450. This equates to a loss of $722 per workday on a relatively simple activity.

References

Alexander & Shankle, Inc., v. Metropolitan Government of Nashville and Davidson County. (2007). Court of appeals of Tennessee, Nashville, TN.

Harmelink, D. J., and Rowings, J. E. (1998). "Linear scheduling model: Development of controlling activity path." *J. Constr. Eng. Manage.*, **124**(4), 263–268.

Hinze, J. W. (2008). *Construction planning and scheduling*, 3rd Ed., Pearson Prentice Hall, Upper Saddle River, NJ.

Thomas, H. R., and Ellis, R. D., Jr. (2007). "Contractor prebid planning principles." *J. Constr. Eng. Manage.*, **133**(4), 542–552.

Thomas, H. R., and Ellis, R. D., Jr. (2009). "Fundamental principles of weather mitigation." *Pract. Period. Struct. Des. Constr.*, **14**(1), 29–35.

Thomas, H. R., Lescher, A., and Bowman, G. (2000). *Fundamentals of contract risk management for electrical contractors*, National Electrical Contractors Association, Bethesda, MD, 67.

[illegible] that the panel [illegible] long [illegible] the [illegible] plant [illegible] to [illegible] [illegible] the [illegible] on a [illegible] simple [illegible]

References

[illegible]

CHAPTER 4

Site Layout

Site layout plans are important because a good plan (1) permits easy movement of materials, equipment, and people around the site; (2) minimizes congestion caused by stored materials and equipment; (3) allows for easy access and retrieval of stored materials; and (4) improves safety. Unfortunately, published literature offers little guidance on how to develop a site layout (Mawdesley et al. 2002; Warszawski and Peled 1987; Zouein and Tommelein 1999). This chapter details a heuristic procedure for developing site layout plans that considers the main elements of site layout planning. If this procedure is followed, the goal of an optimum site layout plan is likely to be achieved.

4.1 Site Optimization

Optimizing the site implies making efficient use of the site for all its intended purposes. Much has been written in the ASCE literature about site layout, but most of the articles describe mathematical modeling techniques that minimize travel distance only. These models tend to be complicated and hard to use, and they fail to address all the aspects important to an effective site layout.

Site layout planning is an important aspect of effective site construction management. Effective site layout contributes to (1) a safer job, (2) reduced job costs, and (3) reduced double-handling of materials. The planner needs to contemplate alternate construction strategies, sequences, and methods. The risks associated with poor site planning are poor housekeeping, congestion, double-handling of materials, inefficient work methods, and reduced safety, just to name a few problems. These are among the more significant and common factors that reduce the efficiency of the labor force (see Fig. 1-1).

Site planning is important for all sites, especially small, constrained ones. With effective site plans, one usually realizes that there is much more space than is initially evident. When combined with many of the principles from the previous chapter, crew idle time can be reduced.

4.2 Procedure

Multiple site plans are needed because the site needs change over time. Normally, three to five plans per project are needed, but for larger, more complex projects, more plans may be justified. Generally, a new plan is needed whenever major material delivery requirements change significantly, such as when moving from concrete work to steel erection; when space or access requirements change; or when new resources, such as a crane, are deleted or added. For a building project, three plans at a minimum are suggested for the three main phases of a project: foundation, superstructure, and service and finish work.

4.3 Developing Site Layout Plans

The steps in developing a site layout plan are given in Table 4-1. These steps should be followed in order because the aspects with the least flexibility or the most importance are addressed first. Fig. 4-1 shows the proposed site plan for the superstructure phase only of the SCMB (see Fig. 3-3 for the actual site layout used). With this proposed plan, drive-through deliveries are possible, and there is more room for on-site storage space if needed. Parking has been moved off site, and the earthen spoil pile is removed. Concrete can be discharged directly from the delivery truck because the site utilities are to be completed early, allowing for this area to be used as a crane pick point (the location of a crane when engaged in lifting) for steel erection. The project office is located on the first floor of the facility once the structural steel activity is completed. If the cranes are sized and located properly, only one pick point per crane will be required. There can be multiple site layout solutions for this phase, and Fig. 4-1 is but one solution.

Table 4-1. Steps in the Development of Site Plans

Step No.	Description
1.	Develop an accurate drawing of the site and surrounding area.
2.	Map E&S facilities and location of drainage facilities.
3.	Locate ingress and egress points for deliveries and pedestrians.
4.	Map the traffic routes.
5.	Locate concrete discharge and crane pick points.
6.	Locate material storage areas.
7.	Map drainage routes (ditches) and locate retention basin.
8.	Locate space for temporary facilities.
9.	Evaluate and refine layout plans.
10.	Repeat steps 2 through 7 for each phase (unique site plan).
11.	Communicate and enforce plans.

Other Site Layout Considerations

It is desirable that there not be drastic changes in the plans from one phase to another. Having few changes means that fewer facilities will have to be moved. Changes to site plans should be planned for a weekend or done after the first shift is over, whenever possible.

For constrained sites, efficient material management is essential (see Chapter 6) because there is minimal room for on-site material storage space. Erecting directly from the delivery truck and preloading are recommended practices for all sites because they will minimize on-site storage needs. Consideration must be given to how to move materials from their on-site stored location or from the delivery truck to the work face.

Key points regarding several of the steps in Table 4-1 are discussed as follows. The steps are then illustrated using three case study examples, which are also described. The one common denominator about the three case study projects is that the sites were small and constrained.

Step 1. Develop an Accurate Drawing of the Site

An accurate drawing of the site is important because cranes and pump trucks need to be accurately positioned to ensure sufficient reach. Accurate locations are important to minimize the number of crane pick points and to ensure the fewest number of relocations. Do not forget important off-site features because this information may be necessary to establish the turning radius of material delivery trucks and other vehicles. The location of overhead utility lines is also important.

Step 2. Map E&S Facilities and Location of Drainage Facilities

An erosion and sediment control (E&S) plan may be required by regulation, but the plan should be expanded to include other drainage issues (see Chapter 5). The drainage of the footprint area and runoff from off-site areas into the site need to be addressed.

In this step, both permanent and semipermanent facilities should be located. Permanently constructed facilities include inlet boxes, drainage pipes, and culverts. Semipermanent facilities include sediment basins and locations of sump pumps. These facilities should be installed early and used as part of the site drainage plan.

Step 3. Locate Ingress and Egress Points

The contractor usually has considerable flexibility in locating site ingress and egress points. However, one needs to also locate building access and points for material deliveries and the removal of trash. Site and building access should be established separately for the workers and delivery trucks. Site craft access, site vehicle access, delivery access, trash removal, and facility access points should all be located to facilitate the easy flow of resources. In general, try to provide for raw material to enter one end of the site and trash to exit the other.

Step 4. Map the Traffic Routes

Using ingress and egress points, separately map pedestrian and vehicular traffic routes. Whenever possible, traffic routes for pedestrians and vehicles should not cross. Trucks delivering materials should not have to back up. Where possible, drive-through deliveries should be planned. A wider than normal gate may be needed in some instances to accommodate the turning radius of longer vehicles.

Step 5. Locate Concrete Discharge and Crane Pick Points

These points vary depending on the method and equipment that are used for concrete placement. The space needs and logistics for discharging directly from the delivery truck, using a crane and bucket, or using a concrete pump truck are uniquely different. In many instances, if the layout plans and construction sequences are done thoughtfully, discharge can be done directly from the delivery truck during the foundation phase. Avoid interferences between discharge and pick points and the traffic routes located in Step 4.

Step 6. Locate Material Storage Areas

If the principles discussed in Chapters 3 and 6 are applied, then the need for on-site material storage is minimal. Do not overlook how deliveries will be moved from on-site storage locations or from the delivery truck into the facility and to the work face (see Chapter 6). Storage areas should be located close to the traffic routes mapped in Step 4 and should provide direct access to facility ingress points.

Step 7. Map Drainage Routes

In this step, drainage routes (shallow ditches) shall be determined. Generally, there is much flexibility in the routing. Try to avoid storage areas, traffic routes, and other sensitive areas. Simple shallow ditches discharging into a small retention basin are sufficient in most instances.

Step 8. Locate Space for Temporary Facilities

Temporary facilities should be located last. Temporary facilities include the office, toolsheds, parking areas, and portable toilets. It may be necessary for workers to park at a remote location and be bused to the site. There may be no room for amenities like on-site parking, spoil piles, or trailers. The temporary facilities should be located as far from the constructed facility as is practical, and consideration should be given to constructing temporary facilities for office space and tools in the constructed facility itself using lumber and polyethylene.

4.4 Case Study 1—State College Municipal Building

This case study project was described in Chapter 3. A proposed layout for the superstructure phase is shown in Fig. 4-1. The site drainage plan is not shown. The project office is a temporary one built on the first floor. Building ingress (for materials), material storage areas, and trash disposal areas are close to the vehicular traffic route. Unimpeded concrete discharge and crane access are provided throughout. Traffic routes do not cross. Crane pick points are close to vehicular traffic routes.

The site area was small, and there was little room for the storage of materials. Site drainage was only a minor issue. Structural steel was the major material being delivered during this phase. Site access and the avoidance of having to back up were major considerations in planning the traffic route for delivery trucks.

4.5 Case Study 2—Bryce Jordan Tower

This case study project is a seven-story apartment building built in State College, Pennsylvania, at a cost of $6.8 million. The building was built in 2003. The ground floor is used for on-grade parking. The superstructure is reinforced masonry, and

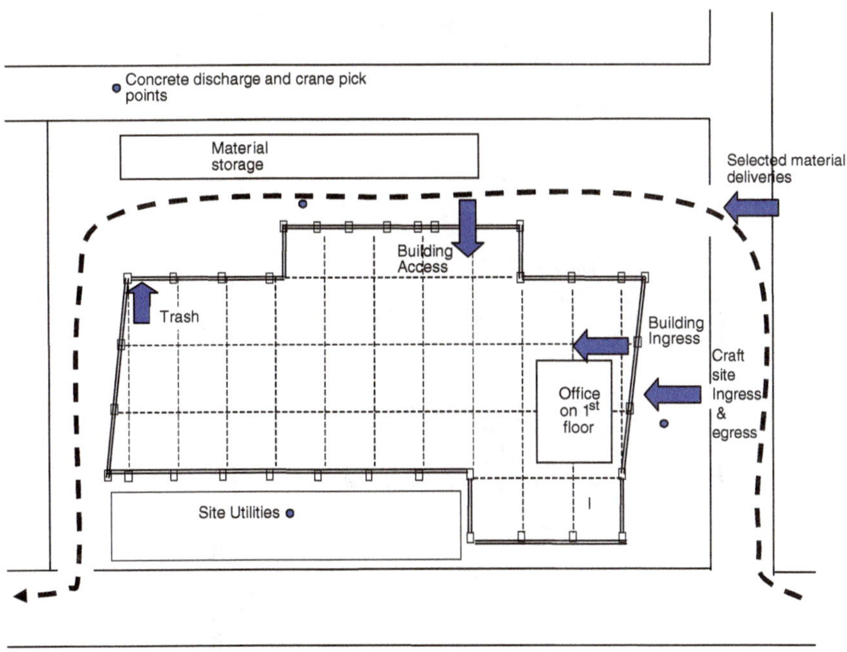

Fig. 4-1. Proposed Site Layout, SCMB

the floors are precast concrete planks. The ground floor and the elevator tower are cast-in-place reinforced concrete.

The main challenge in developing a site plan was that the site area was small and confined. The building footprint occupied most of the site area, meaning that there was little room for material storage. For the superstructure phase, many material deliveries were required.

The proposed site plan for the superstructure phase consists of two delivery routes, as shown in Fig. 4-2. The site drainage was inconsequential, and a plan is not

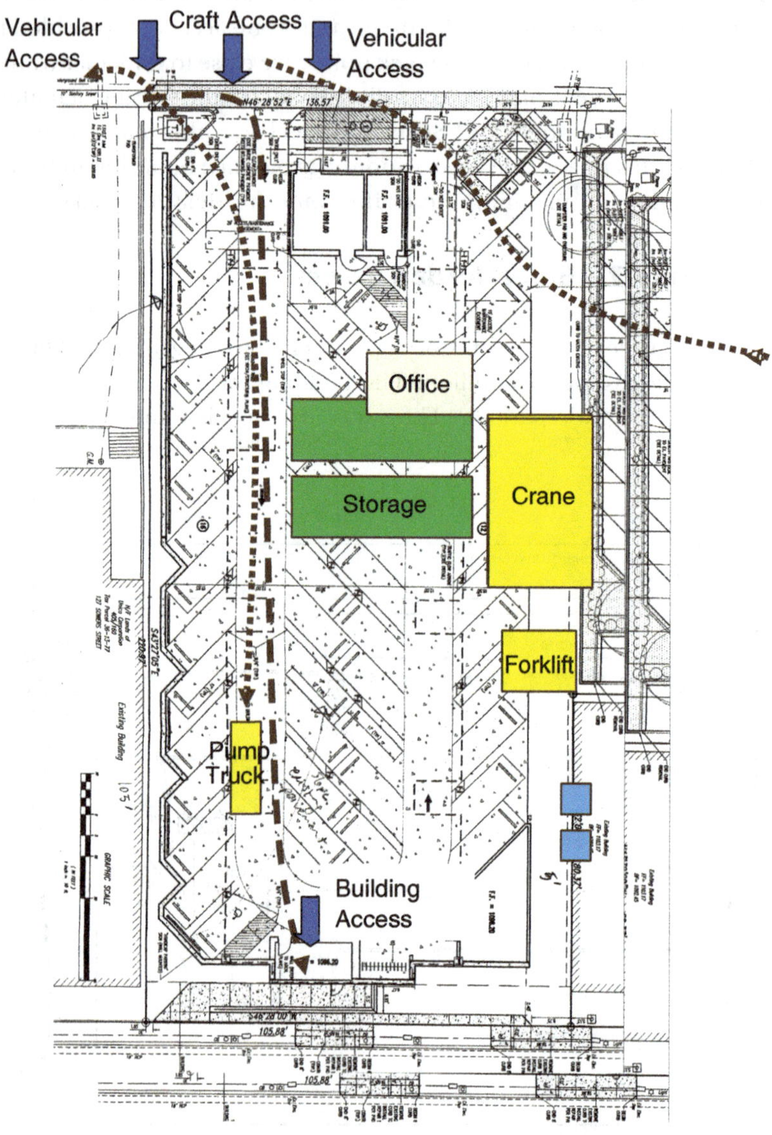

Fig. 4-2. Proposed Site Plan for Superstructure Phase of the Bryce Jordan Tower

shown here. The major materials being delivered during the superstructure phase were concrete block, steel reinforcement, grout, precast floor planks, and concrete. With this plan, selected bulk materials (masonry and floor planks) were delivered to one side of the building (project east), and concrete and grout were delivered to the opposite side of the building (project west). The project office was built on the second floor. Traffic routes did not cross. There was no on-site parking.

4.6 Case Study 3—Beaver Avenue Parking Garage

The Beaver Avenue Parking Garage was built in State College, Pennsylvania, in 2007 at an approximate cost of $11 million. The garage is a seven-story precast concrete structure. The precast erection was divided into five phases. A cast-in-place concrete duct bank ran through the middle of the project. Phases 1–2 of the precast erection work were on the south side of the duct bank. Phases 3–5 were on the north side. The plan was that the duct bank was to be finished before the final three phases of the precast erection (project north) could begin. Fig. 4-3 shows the actual site layout

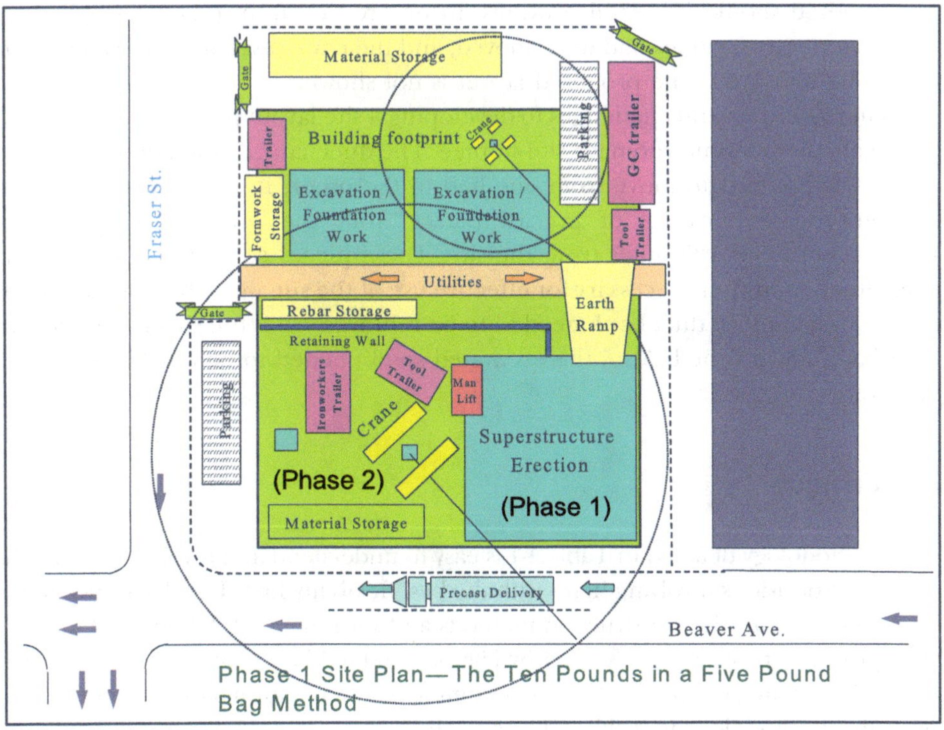

Fig. 4-3. Actual Site Layout for Phase 1 of the Beaver Avenue Parking Garage

Fig. 4-4. Actual Site Conditions, Phase 2, Beaver Avenue Parking Garage

used by the contractor for phase 1 of the superstructure. The contractor used the "ten pounds in a five-pound bag" method of site layout. Fig. 4-4 shows the actual site conditions during phase 2. The consequences of the "ten pounds in a five-pound bag" method should be readily obvious. In progressing from phase 1 to phase 2, many temporary facilities had to be moved, and the crane used for precast erection had to be relocated. The proposed layout is not shown.

There were several challenges to developing a site layout plan. Because the site was small, there was no room for site storage. The duct bank divided the site in half, and it was not easy to move from one half to the other. It was necessary to have frequent daily deliveries of precast pieces from the precast facility, which was located about 125 miles from the site. These pieces could not be stored on site. Other conditions necessary for effective use of the site were that the formwork for the cast-in-place duct bank needed to be removed from the site as soon as the duct bank was finished. Also, the excavated spoil needed to be quickly reused or removed.

4.7 Critique

The methodology detailed in Table 4-1 is easy to understand and involves making the important decisions involving items with the best flexibility first. The decisions should be integrated with the construction methods and with material delivery and general strategies. The projects represented by Figs. 4-1 and 4-2 had much more open space than initially envisioned and did not become congested with materials and facilities once the methodology in Table 4-1 was applied.

References

Mawdesley, M. J., Al-Jibouri, S. H., and Yang, H. (2002). "Generic algorithms for construction site layout in project planning." *J. Constr. Eng. Manage.*, **129**(5), 418–426.

Warszawski, A., and Peled, N. (1987). "An expert system for crane selection and location." *Proc., 4th Int. Symp. on Robotics and Artificial Intelligence in Building Construction*, Vol. **1**, Israel Institute of Technology and Building Research Station–Technion, Haifa, Israel, 64–68.

Zouein, P. P., and Tommelein, I. D. (1999). "Dynamic planning using a hybrid incremental solution method." *J. Constr. Eng. Manage.*, **125**(6), 400–408.

PART III

Management Factors that Lead to Improved Productivity

CHAPTER 5

Fundamental Principles of Weather Mitigation

Adverse weather is an unavoidable fact of life for construction contractors. The problem is more pronounced as owners want project completion in a shorter time frame, meaning that contractors cannot easily avoid working during periods of adverse weather. Often, contractors are called upon to work soon after a weather event, such as a rain or snowstorm, has occurred. The effects of the event may still be present.

In general, contractors seem ill-prepared to respond to adverse weather, and the literature (see Chapter 2), continuing education courses, and the Internet offer little or no guidance about how to mitigate the negative effects to cost and schedule. Adverse weather can lead to serious negative effects on labor and equipment productivity. The problem addressed in this chapter is how the negative effects of adverse weather can be avoided or minimized so that the workers can return to work quickly while avoiding the need for rework or working in a less than ideal environment (Thomas and Ellis 2009). Without effective weather planning, there is a risk that labor and equipment will need to work in unfavorable and inefficient conditions.

5.1 Weather Effects

The effects of adverse weather can arise in a few ways: (1) extreme temperatures and humidity and (2) weather events. Temperature extremes can be day to day or seasonal. Weather events include rain, snow, wind, and ice (Clapp 1966). The effects of adverse weather events can last long after the event itself occurs.

On three construction activities that were observed by the authors, the labor efficiencies from adverse weather were quantified as 22%, 8%, and 27% of the normal output. The actual productivity values compared to normal were 78%, 92%, and 73%. The lingering effects from mud, rainfall runoff, and snow have been observed on many other projects. Considering the cost of labor, almost any weather mitigation strategy is cost effective. In this chapter, relatively simple and inexpensive strategies are described to lessen or eliminate the negative effects.

There are several added benefits that can accrue from effective weather mitigation strategies. The project is likely to be safer, and, in buildings, there is a reduced likelihood of mold or water damage, which can lead to rework.

5.2 Estimating Example

A simple estimating example shows the importance of considering weather effects. It is based on the probability of a particular weather event occurring and the estimated average effect (see Table 2-2) resulting from that event. Cold and hot temperatures are considered. The probability of a weather event and cold and hot temperatures occurring derives from local weather records and is a historical value. The probabilities also depend on the planning horizon used. For a one- to two-year project, a planning horizon of 10 years seems reasonable. The effects from humidity and wind are not considered in this example. Productivity inefficiency percentages unique to a particular activity can be substituted for the estimated average values from Table 2-2 because effects can vary appreciably. However, it is important that average values be used and that contractors state their assumptions at the time of the bid submission. Impacts should be applied only to activities that may be affected. The example is simplistic. It is unrealistic to assume that all construction activities will suffer the same degree of productivity loss because of the same set of weather conditions.

Suppose it has been estimated that an activity that will last for 12 months will take 7,600 work hours to complete under ideal weather conditions. Assume that by consulting the weather data for the locale where the project is to be built (based on a 10-year planning horizon), the following probabilities and effects are determined:

- Probability of cold temperatures (<20°F) is 0.01 (impact = 50%),
- Probability of hot temperatures (>85°F) is 0.02 (impact = 60%),
- Probability of rain is 0.01 (impact = 40%), and
- Probability of snow is 0.01 (impact = 50%).

The activity is fully exposed to the weather. How many work hours should be added in the estimate to account for adverse weather? The work hour estimate is determined as follows:

Consider the hot temperature effects. Weather data indicate that for a 10-year planning horizon, there is a 0.02 likelihood that work will be performed in temperatures greater than 85°F. Whenever work is performed under these conditions, the output of the crew is reduced to 60% of the normal daily output that would be achieved during favorable weather. Thus, the loss of efficiency is 40%, and the impact factor is calculated as 1.00/0.40 = or 1.67. But productivity losses only occur 2% of the time. The labor effects in terms of work hours is the following:

$$\text{Additional Labor Hours Due to Hot Weather} = (7{,}600)\ \frac{1.00}{0.60}\ 0.02 = 253$$

And for all weather effects combined is the following:

$$\begin{aligned}\text{Work Hour Estimate} &= 7{,}600 \times \left[1.00 + \left(\frac{1.0}{0.50}\right)0.017\left(\frac{1.0}{0.6}\right)0.02 + \left(\frac{1.0}{0.7}\right)0.01 + \left(\frac{1.0}{0.5}\right)0.01\right] \\ &= 7{,}600 \times (1.098) \\ &= 8{,}347\end{aligned}$$

The work hour estimate in this example should be increased by 0.098%. The first perception is that this percentage is somewhat high, but one should realize that the reduced output per day when adverse weather is present is significant (see Table 2-2) and the output with which this value is being compared is the normal output during ideal weather. The probabilities of occurrence have been assumed. The economic effect (at $35/h) is almost $30,000. Thus, any easily implemented, low-cost strategy that will reduce this dollar amount seems justified.

5.3 Fundamental Principles

The analysis of the literature highlights the need for contractors to plan to mitigate the effects of adverse weather. Sadly, many contractors do an inadequate job of planning to avoid the negative consequences of adverse weather. Table 5-1 lists inexpensive fundamental principles for avoiding the negative effects of adverse weather. If these principles are followed, contractors can avoid losing large sums of money because of adverse weather. The principles are organized into four categories and are applicable to a wide range of projects and conditions.

General

Table 5-1 lists five general principles that relate largely to planning. Using a 4-10 schedule (four 10-h workdays) (Principle 1.1) allows Fridays to be used as a makeup day for lost time without having to pay overtime wages (provided that the workweek does not exceed 40 h). Though there has been little research into the effects of a 10-h workday, all indications are that a 10-h workday is not detrimental to labor productivity compared with an eight-hour workday. The 4-10 schedule reduces the number of daily startups and shutdowns and reduces the contractor's craft travel time and expenses. Workers like a 4–10 schedule.

Using annual cycles to schedule around trades that are most susceptible to adverse weather (Principle 1.2) may not always be feasible, except on projects with lengthy schedules and ones with considerable schedule flexibility. Accelerating a schedule to avoid winter weather (Principle 1.3) must be done with care, and

Table 5-1. Fundamental Principles of Weather Mitigation

No.	Principle
General	
1.1	Use a 4-10 work schedule to permit a makeup day on straight time.
1.2	Use annual cycles to schedule around trades most affected by weather.
1.3	Where justified, accelerate the schedule to avoid winter cost.
1.4	Where possible, reserve some work that can be done on inclement workdays.
1.5	Enclose and seal buildings for weather protection.
Excavation and Site Work	
2.1	Seal exposed areas each day.
2.2	Use drainage systems and sump pumps and plan for runoff, especially from off site.
2.3	Make use of permanent site drainage by installing drainage facilities early.
2.4	Apply an adequately engineered working surface.
2.5	Plow wet ground to accelerate drying.
2.6	Build and maintain all-weather roads.
2.7	Leave snow in place as insulation until ready to excavate.
Labor	
3.1	Shift work hours to avoid the heat of the day.
3.2	Provide break trailers for relief from cold or heat.
Materials	
4.1	Protect materials from weather.
4.2	Store materials on timbers and pallets to keep them out of mud.

an economic analysis may be justified. Accelerating limited items of the work is probably sufficient; it is probably not necessary to accelerate the entire job.

Fig. 5-1 shows the status of the masonry on the Oliver Middle School in February 2004 (see Chapter 3). By this time, it was obvious that the project would not be completed on time and that schedule acceleration was necessary. This fact was probably foreseeable months sooner. Because masonry was a substantial part of the work, accelerating masonry alone should have been contemplated. Yet there is no

Fig. 5-1. Oliver Elementary School, February 16, 2004

evidence of acceleration in Fig. 5-1 because minimal materials are observed. One strategy to accelerate the masonry would be to build with concrete masonry units and brick concurrently or to use multiple workstations. But it would be necessary first to remove the mud and to compact the area in the forefront of the figure.

Reserving work for inclement workdays (Principle 1.4) carries certain risks. It may not be possible to apply this principle on many projects. On one project where a dam was being constructed, it was observed that certain work was reserved for inclement workdays, and then on inclement weather workdays, the workers were sent home. It was little wonder that the project schedule slipped. Maybe a different strategy should have been adopted.

On many commercial projects, completion of the roof, allowing for the enclosure of the facility, is often treated as an important milestone. Roof completion is often an "essential to success (ETS)" sequence, and a concerted effort may be justified to reach this milestone.

Principle 1.5 deals with sealing a building. Sealing a building from rain is a separate, but equally important, function from enclosing a building. Enclosure is associated with winter activities, whereas sealing is often associated with springtime or rainy weather. Preventing water infiltration is important to preventing mold, which leads to rework. Sealing a building from rain and enclosing a building to allow temporary heat are two different actions, but both can be accomplished simultaneously with permanent windows or temporary enclosures. Permanent windows are preferred to temporary enclosures, but windows may be a long lead item that requires early submittal approval. There may be cost implications. It has been observed that contractors often do a poor job with temporary building enclosures. Wind damage to the enclosure often necessitates rework, at considerable expense to a contractor. Fig. 5-2 shows an example where rework may cost the contractor thousands of dollars in labor cost to repair. The concern for mold alone justifies permanent windows, and there will likely be cost/rework issues. A temporary enclosure using polyethylene is often a marginally effective choice for enclosing

Fig. 5-2. Temporary Building Enclosure Damaged by Wind

and sealing work areas, and many situations such as the one shown in Fig. 5-2 have been observed. Using permanent windows largely eliminates subcontractor "callback" time and improves the wintertime work environment for the crafts.

Excavation and Site Work

Earthwork is particularly susceptible to weather effects, and thus, Principle 2.1 should be considered. Even on a building project, sealing exposed areas (Principle 2.1) is recommended. But sadly, a compactor is rarely seen or used on a commercial building project. Figs. 5-3 and 5-4 show the potential consequences of not sealing.

Fig. 5-3. Muddy Site Conditions Caused by Rain and Failure to Seal Exposed Areas

Fig. 5-4. Muddy Site Conditions Caused by Snow and Failure to Seal Exposed Areas

One can only imagine the amount of additional labor effort needed to contend with these muddy conditions. This principle can greatly improve working conditions.

Planning and implementing site drainage management is an important factor in maintaining a workable site. Permanent and temporary drainage facilities should be used to plan for and control runoff. Do not ignore runoff from off-site sources (Principles 2.2 and 2.3). Permanent drainage should be installed as early as practical (Principle 2.3). On one project, it was observed that the contractor installed the permanent drainage at the end of the as-built schedule. The superintendent claimed that he did not want to drive delivery trucks over the newly installed facilities. He also claimed that there was never a problem with runoff, despite the fact that on several occasions, the ponding of a foot or more of water was observed exactly where the permanent facilities were to be located. Fig. 5-5 shows an example where runoff from off site was not channeled away from the site. This oversight cost the contractor thousands of dollars. On a commercial site, channel rain to the lowest area of the facility footprint and use a sump pump to expel the water from the site. Fig. 5-6 shows an example where this method was used effectively. The lowest point on the site is seen in the lower left portion of the photo.

A site drainage plan consists of three parts: (1) drainage of the facility footprint, (2) drainage of the remainder of the site, and (3) measures to prevent off-site runoff from flowing into the footprint area or onto the site. Each of these aspects can be addressed in most cases with shallow ditches and sump pumps. A simple but effective drainage plan, as shown in Fig. 5-7, should be integrated with an erosion and sediment control (E&S) plan. All three components of the plan can be seen in Fig. 5-7. The runoff drains to the lower right of the figure, which is the lowest elevation of the site. The application of Principles 2.2 and 2.3 should channel rainfall away from critical areas of the site, greatly reducing muddy working conditions.

Fig. 5-5. Flooded Footers Caused by Runoff from Off Site

Fig. 5-6. Drainage of Facility Footprint

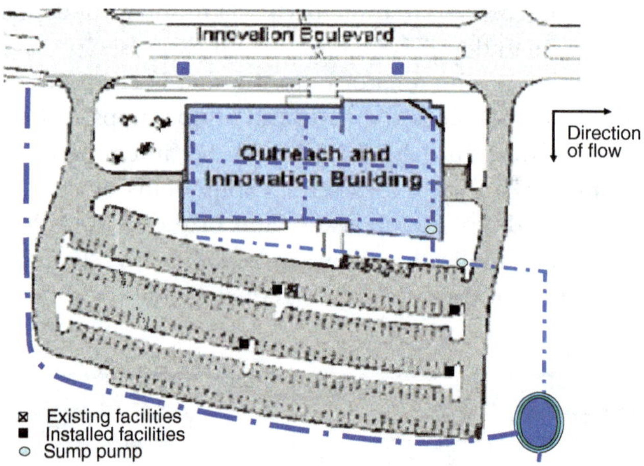

Fig. 5-7. Site Drainage Plan, Outreach Building Project

An engineered working surface should be applied to prevent muddy conditions (Principle 2.4). It should be designed using sound engineering principles and should be properly compacted and constructed to provide stable support for crane outriggers. Unfortunately, many professionals think all that is needed is to place some gravel on the site. Gravel is placed only when muddy conditions are evident. These surfaces sometimes fail, as shown in Fig. 5-8.

Another option is to consider the use of more weather-resistant materials. For example, asphalt concrete pavement base material may be substituted for certain rock pavement base materials. Likewise, when rainy conditions are anticipated, the selection of a clean, free-draining fill material may mitigate weather effects.

Fig. 5-8. Failed Site Area

Other helpful principles that are largely applicable to earthwork projects are to plow wet ground to accelerate drying (Principle 2.5) and to build and maintain all-weather roads (Principle 2.6). Roadside berms and ledges should be flattened to allow water to drain away from the haul road (Fig. 5-9). Unless berms are flattened, rainfall will not drain and muddy areas will develop. In winter work, leave snow in place as insulation until the contractor is ready to excavate (Principle 2.7).

The cost implications in this category arise from the use of a compactor and grader and the application of an engineered working surface. However, one must balance these additions to the work scope against the cost of more labor (arising from inefficiencies) or the cost of a crane accident if the working surface fails.

Labor

To avoid work in extreme hot or cold temperatures, shift the work hours earlier to avoid the heat of the day (Principle 3.1) and provide break trailers for relief from the heat and cold (Principle 3.2). Shifting work hours can also be done to increase the hours of daylight or to avoid heavy city traffic.

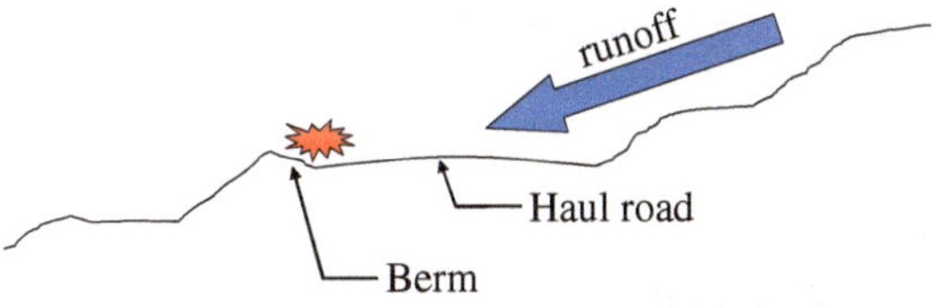

Fig. 5-9. Illustration of Berm Problem that Needs to Be Corrected

Fig. 5-10. Proper Storage Practices to Protect Materials from the Weather

Materials

Material management is important to managing a cost-effective or efficient site (see Chapter 6). Certain materials need to be protected from the weather (Principle 4.1). Fig. 5-10 shows effective storage practices from one project. A working surface has been applied (Principle 2.4), the materials are stored on pallets (Principle 4.2), and materials are protected (Principle 4.1) from rain and snow. The importance of storing materials on timbers or pallets (Principle 4.2) is twofold. Storage off the ground provides protection from mud and standing water, and retrieval is easier if a forklift or other lifting device is used. The cost implications in this category are minimal.

5.4 Developing Weather Mitigation Plans

Developing weather mitigation plans is an important component in the planning process. The E&S plan (if required) should be augmented with site drainage plans and enclosing and/or sealing strategies. While weather mitigation plans can vary, the following components are essential parts of a comprehensive plan.

Contents of Weather Mitigation Plans

Facilities

- Existing facilities, e.g., inlet boxes and culverts;
- Permanent facilities to be installed, e.g., inlet boxes, culverts, drainage pipe, and swales; and
- Temporary facilities, e.g., retention basins, straw bales, and silt fences.

Drainage

- Drainage routes,
- Sump pump locations, and
- Retention ponds.

Many of the elements cited here are covered in E&S plans. Other elements of the plan should include how the building will be enclosed and sealed, the work schedule, how crafts will be protected from temperature extremes, and other important aspects of weather mitigation.

Procedure

The procedure for developing weather mitigation plans is given in Table 5-2. The procedure involves three broad components. The first is to locate existing permanent and temporary facilities. Facilities include inlet boxes, culverts, and drainage pipes. Next the drainage routes are mapped. Finally, other decisions are made, e.g., about how and when the building will be enclosed and sealed and about the work schedule.

Key points regarding several of the steps in Table 5-2 are discussed as follows using the site plan for the Outreach Building shown in Fig. 5-7 (see also Fig. 3-7). The drainage plan needs to be developed in conjunction with the site layout plan described in Chapter 4.

1. Develop an Accurate Drawing of the Site—An accurate plan needs to be developed. The site depicted in Fig. 5-7 is relatively flat. The direction of the drainage is shown. The lowest point on the site is the southwestern corner. Elevations, as appropriate, need to be known.

2. Locate Existing Drainage Facilities—The existing facilities include inlet boxes, culverts, stormwater drains, and swales. Four existing inlet boxes are shown in Fig. 5-7. One should be sensitive to a situation that some older existing systems may not be properly sized.

3. Locate Permanent Drainage Facilities that Are to Be Installed—One also needs to identify those facilities that will be used during construction. Some or all facilities may not be usable. Fig. 5-7 shows four new inlet boxes in the parking lot area. The installed facilities cannot be used until the parking lot has been paved.

Table 5-2. Procedure for Developing a Weather Mitigation Plan

No.	Description
1.	Develop an accurate drawing of the site.
2.	Locate existing drainage facilities.
3.	Locate permanent drainage facilities that are to be installed and identify those that will be used.
4.	Locate discharge points, retention ponds, and temporary facilities.
5.	Map on-site drainage routes.
6.	Map off-site water intrusion controls.
7.	Identify sump pump locations.
8.	Establish strategies for facility enclosure and sealing.
9.	Determine schedule considerations.

4. Locate Discharge Points, Retention Ponds, and Temporary Facilities—The site discharge point and a small retention pond are located at the southeast corner of Fig. 5-7.

5. Map On-Site Drainage Routes—Drainage routes are usually shallow ditches. There are two areas of concern: (1) the facility footprint and (2) other site areas. Both areas are shown in Fig. 5-7. Locating drainage routes should be one of the last steps in developing a site layout plan (see Chapter 4). Routes should be located to avoid such things as high-traffic areas and material storage locations.

6. Map Off-Site Water Intrusion Controls—Do not forget about rainfall entering the site from off-site areas. A small ditch can usually alleviate the problem. In Fig. 5-7, this issue was not a major concern. But any rainfall draining to the site from the north will be handled by the two existing inlet boxes located in the street (north), and water that may enter from the east is diverted by a perimeter ditch.

7. Identify Sump Pump Locations—Once the rainfall has been channeled from the work area and site to a holding area, it may be necessary to use sump pumps to convey the water to a higher elevation or to negotiate travel routes (pedestrian and vehicular). In Fig. 5-7, two sump pump locations are identified to remove rainfall from the building footprint.

8. Establish Strategies for Facility Enclosure and Sealing—The facility needs to be both enclosed (for heating) and sealed (to prevent water intrusion). There are two options: temporary enclosures and permanent windows. Permanent windows are the preferred option, but long lead times may preclude this option. On the Outreach Building, permanent windows were used.

9. Determine Schedule Considerations—Installing permanent windows may require changes to ordinary schedule logic. Consider the Outreach Building shown in Fig. 3-7. From the orientation shown, the building is divided into two halves; the right half is built first, followed by the left half. Starting masonry early on the right half means that the glass curtain wall has a good chance of being completed before building enclosure is required. Also, on the Outreach Building project, the parking lot needs to paved before the permanent drainage facilities can be used.

5.5 Case Study 1—Painted Post, New York, State Route 17 Interchange

A case study project is used to illustrate the effects of adverse weather on labor productivity on a single project. This case study also illustrates the variety of ways construction operations can be affected.

This contractor did not use a 4-10 work schedule (Principle 1.1). Also the following principles from Table 5-1 are not applicable to this project: 1.2, 1.3, 1.5, 2.4, 2.7, 3.1, and 3.2. It is not known if the following principles were applied: 1.4 and 2.5. The following principles were applied where applicable: 2.1, 2.2, 2.3, 2.6, 4.1, and 4.2.

Activity Descriptions

Weather is an issue that affects many types of construction operations in unique ways. Highway work is particularly susceptible to the adverse influence of unfavorable weather because most highway operations are fully exposed to the weather, and some materials can be highly susceptible to being affected by adverse weather. Bridge construction work is also vulnerable to the effects of adverse weather. The cases in which these types of work become isolated from the weather are rare, and weather can often have a significant effect on a project for its entire duration. This case study examines the effects of weather on a single highway project and how adverse weather affected labor productivity uniquely for different activities.

The case study project is the construction of a highway interchange on Route 17 in southern New York state that was built by a local nonunion contractor. The seven activities studied were chosen because they ordinarily required varying levels of manual labor, and the productivity on each activity was affected differently by the weather. The activities studied include the driving of three different types of piling, earthmoving from a borrow site to a fill site using trucks, concrete pavement repair by dowel bar retrofit, concrete pavement production, and asphalt pavement production. Production data were obtained from daily time sheet summaries. Notes about weather conditions and their effects were made on these sheets as well. Discussions were also conducted with superintendents and supervisors about how particular weather events affected the conduct of the operation. Some information was also obtained from daily logbooks kept by superintendents. All of these data were then compared with weather data for the days that the work was performed.

Weather effects on work operations on this project were observed in two categories. The first category included work stoppages mainly because of safety concerns, lack of access to the site, or concerns over a detrimental effect on the quality of the constructed product. The second category was decreased efficiency and loss of productivity because of weather conditions while the work continued. This resulted in worker discomfort, trucking delays, time spent preparing the site for work because of weather-related issues, and a variety of other reasons having to do with the weather.

Sheet-Pile Driving—Sheet-pile driving entails hooking a long, corrugated sheet of steel onto a vibratory hammer hanging from a crane and allowing the vibrations of the hammer to drive the sheet into the ground. Adjacent sheets are interlocked as they are driven and are used to create cofferdams to build structures in areas that were sometimes covered with water. They may also be used to create retaining walls so that excavation adjacent to a pile wall may be carried out.

The crew for driving sheet piles generally consisted of a crane operator, a worker who directed the crane operator and operated the pile hammer, a laborer who aligned the sheets and hooked them to the leads and the pile hammer, and a welder who prepared sheets for driving by splicing them or attaching lifting points to the

Fig. 5-11. Sheet Pile Cell

sheets and cutting the excess steel from sheets were already driven. A typical completed sheet pile operation is shown in Fig. 5-11.

H-Pile and Tube-Pile Driving—The methods used for driving H-pile and tube piles are similar to that used for sheet piles. These piles are load-bearing piles used in the bridge foundation. The piles are usually driven with a diesel-operated pile hammer instead of a vibratory hammer. The crew size is the same as for sheet-pile driving. A typical H-pile operation is shown in Fig. 5-12.

Earthwork—The earthmoving operations observed on the case study project were varied, including the conditions under which the work was done, the length of the haul, and the equipment moving the earth. The earthmoving operations were

Fig. 5-12. H-Pile Driving

optimized such that a truck was always being loaded. All the earthwork operations reported were trucks being loaded with excavators. All-weather roads were constructed and maintained (Principle 2.6, Table 5-1). The crew for the earthwork operations consisted of a supervisor and operators for each piece of equipment. Grading at the cut and the fill was performed by track-type tractors equipped with a global positioning system so that grade checking or staking was never a limiting factor. Some work locations were beside a river, so the worksite was vulnerable to flooding, as shown in Figs. 5-13 to 5-15.

Dowel Bar Retrofit Concrete Pavement Repair—Dowel bar retrofit is a method of repairing concrete pavement slabs that involves sawcutting slots across a crack in the concrete, chipping and cleaning the slots, placing dowel bars, and filling the slots

Fig. 5-13. Earthwork (Vulnerable to Flooding)

Fig. 5-14. Pile Driving in Flooded Conditions

Fig. 5-15. The Effect of Flooding on a Construction Project

with concrete. This is a labor-intensive operation, and most of the work is performed manually. Sawing of the slots was not included in the study because all of the sawing was done ahead of time by a subcontractor. The crew for this operation consisted of two chisel operators for removing the concrete; two laborers cleaning the slots; one worker placing the dowels; and three workers mixing, placing, and finishing the concrete to fill the slots. Fig. 5-16 shows a typical operation.

Concrete Paving—Concrete paving was the most comprehensive and personnel-intensive operation observed. It involved the placement of new concrete pavement with a slipform paving machine. Concrete was supplied by an on-site concrete batch

Fig. 5-16. Dowel Bar Retrofit, Concrete Pavement Repair

plant, which was not part of the study. The crew for this operation consisted of four operators for equipment to place the concrete, four finishers, one laborer to direct the trucks, a supervisor, and drivers for each of the concrete trucks. Thus, the crew was a mix of laborers and operators who were fully exposed to the weather.

Concrete paving itself is a very weather-sensitive operation because even a small amount of rain can dramatically affect the contractor's ability to produce high-quality pavement. Because the influence of weather is so great on concrete paving operations, these operations did not occur every day. They are generally planned well in advance and in coordination with the weather forecast. If there was a day that paving could not be done, the paving crew was reassigned to another activity. Only days when paving occurred are included in the data. Fig. 5-17 shows a typical concrete paving operation.

Asphaltic Concrete Paving—Asphaltic paving involves the laying of asphaltic concrete pavement. Asphaltic concrete was supplied from an off-site plant via dump trucks. The data chosen were the closest to having the right number of trucks delivering asphaltic concrete to keep the paver running continuously. This method served to eliminate losses in productivity caused by lack of materials. A paving crew generally consisted of a supervisor, two laborers to rake the asphaltic concrete, a paver operator, four compactor operators, and truck drivers delivering the asphaltic concrete. Asphalt paving is an intense operation because of the heat to which the workers are exposed. The asphaltic concrete is delivered at about 300°F, and the workers are exposed to this heat almost continuously. Much of the manual labor has been taken out of this operation with the development of better pavers and the use of skid-steer loaders for placing asphaltic concrete in areas where the paver cannot reach. Fig. 5-18 shows a typical asphaltic concrete paving fleet.

Fig. 5-17. Concrete Paving Operation

Fig. 5-18. Asphalt Paving Operation

Weather Data

All weather data were provided by the National Weather Service. Several parameters were developed to allow easier correlations to construction data. The weather events noted were when (1) there was rain on that workday, (2) there was substantial rainfall in the preceding three days, (3) there was thunderstorm activity in the area that day, (4) winds exceeded 25 miles per hour that day, and (5) the high temperature that day exceeded 85°F. Because this study was only during periods of warm weather work (late spring and early summer), only days that were exceptionally hot are contained in the study. Thus, the case study does not encompass cold temperatures. Temperatures during the case study generally were in the range of 68–89°F.

Results

All five categories of weather events were documented. These were the following:

- Winds in excess of 25 miles per hour,
- Thunderstorms in the area (not necessarily at the site),
- Significant rain the previous three days,
- Rainfall that workday, and
- High temperature exceeding 85°F.

Sheet-Pile Driving—In the case of sheet-pile driving, there was no visible effect on productivity because of high temperatures. This fact is not surprising because this operation is not particularly labor intensive. Despite the fact that three of the four crew members were exposed to the elements, the work is largely equipment intensive. Pile driving does involve long steel members and crane booms at significant heights. For this reason, the threat of thunderstorms and high winds can have a greater effect on productivity than does temperature. Therefore, when these conditions were imminent, the operation was often suspended for safety reasons. In addition, these operations were often carried out in low-lying areas, so the operation was susceptible to flooding and rain. Flooding can either impede access to the worksite, slow the flow of materials and equipment, or shut the work area down completely. Principle 2.2 was applied to this activity.

Fig. 5-19 shows the productivity of the sheet pile operation. The operation was significantly affected by thunderstorms (workdays 11–13 and 19–22), but the work was also affected by rain (workdays 1–2 and 21–25) and high winds (workdays 21 and 23). The last week of observation saw temperatures in excess of 85°F. There was some effect, but the effect was minimal compared with the weather events. Overall, there were few disruptions and the operation seemed to consistently achieve a sustainable productivity of about 0.23 work hours/m. Weather played a significant role in making the cumulative productivity 0.31 work hours/m (about a 34% increase). Thunderstorms seemed to have the greatest effect.

H-Pile and Tube-Pile Driving—Tube-pile and H-pile driving showed effects and vulnerabilities to the weather similar to sheet-pile driving. Again, temperature did not seem to have a major effect on productivity, but potential thunderstorm activity and rain did. Flooding of the work area was not quite as significant an issue in these operations compared with sheet-pile driving because the areas where these piles were being driven were often already protected from flooding by the sheet-pile walls and cofferdams. Also, one work area of H-pile driving that was studied was in a well-drained area that never saw significant rain collection or ponding. However, the crew productivity was somewhat affected by the rain. Overall thunderstorms in the area seemed to have the greatest effect, and high temperatures the least. The cumulative

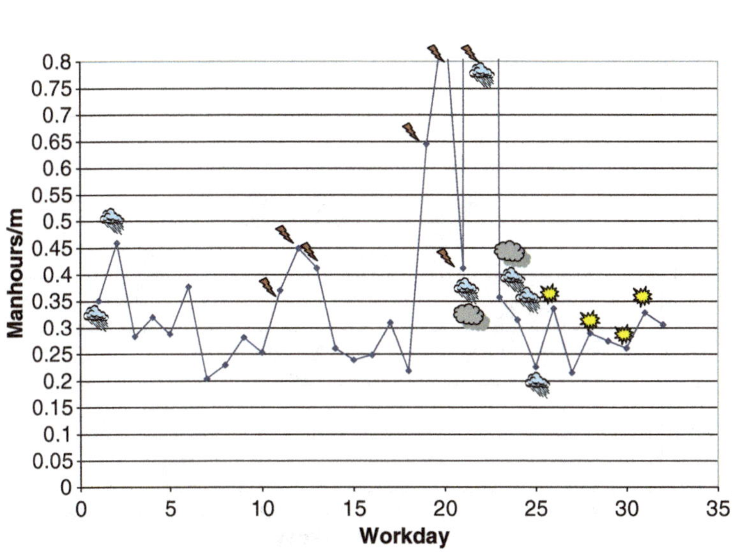

Fig. 5-19. Sheet Pile Productivity

productivity of the H-piling and tube-piling operations was 0.41 work hours/m (H-piling) and 0.37 work hours/m (tube piling). The actual productivities of H-piling and tube piling are shown in Figs. 5-20 and 5-21, respectively. Generally speaking, in the absence of weather disruptions, the sustained productivity rate of all three piling operations was in the same range, 0.25–0.30 work hours/m.

Earthwork—As seen in Fig. 5-22, earthwork was also not significantly affected by high temperatures. It was, however, greatly affected by rainfall and flooding problems. The sites studied in this situation were somewhat unique in that there were areas that

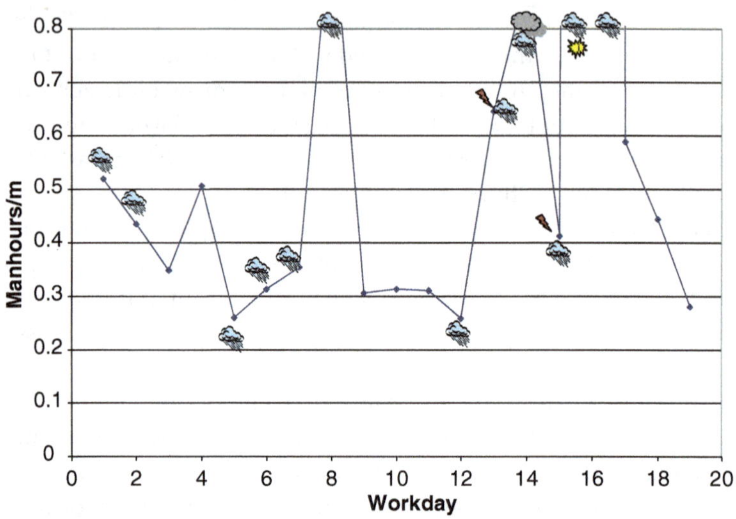

Fig. 5-20. H-Pile Productivity

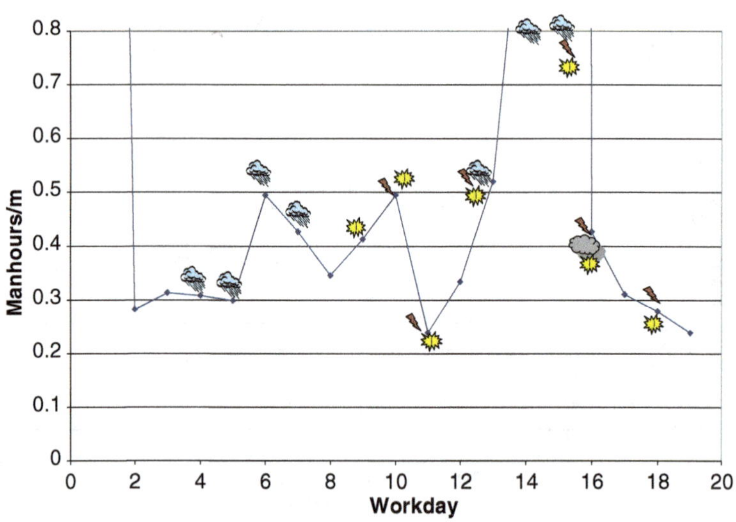

Fig. 5-21. Tube-Pile Productivity

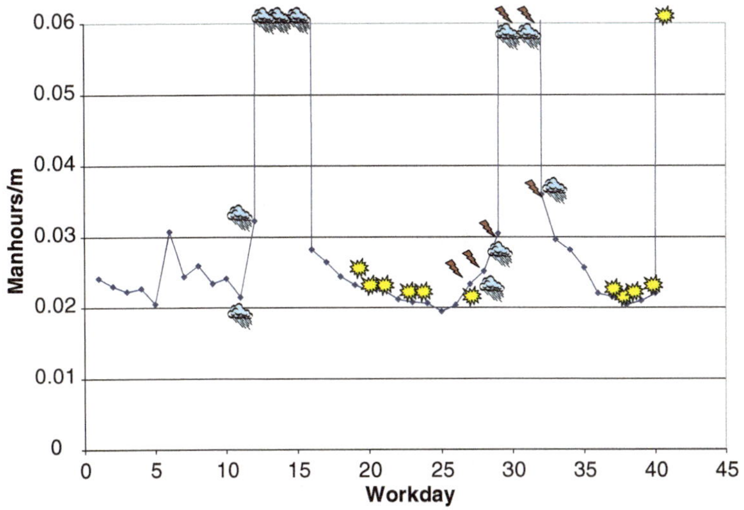

Fig. 5-22. Earthwork Productivity

were vulnerable to flooding, yet some areas dried quickly, and there were other less vulnerable sites that could be used when flooding occurred. The proximity to a flood-prone river often necessitated these adjustments, and the ability to change locations seemed to provide a mitigating effect on the productivity of this operation. Thus, providing alternate worksites, where possible, is important (Principle 1.4). Many of the fundamental principles in category 2 of Table 5-1 were applied to this activity.

The effect of sustained rainfall (see Fig. 5-22) occurred around workdays 15 and 32. In both instances, it took 5–10 workdays for the material to dry sufficiently for the operation to return to normal. It is not known if Principle 2.5 was applied. Thunderstorms seemed to have little effect unless it actually rained at the site. Overall, the sustained productivity of this operation was about 0.009 work hours/m^3. The cumulative productivity was calculated as 0.0108 work hours/m^3 (21% increase).

Dowel Bar Retrofit Concrete Pavement Repair—The dowel bar retrofit seemed to show little vulnerability to high temperatures. Because this activity occurred on existing pavement that was well drained, rainfall did not directly affect dowel bar retrofit unless the rainfall became heavy. There was a secondary effect of rainfall, though, in that some of the work occurred adjacent to an active expressway. If the weather conditions resulted in decreased visibility of the motoring public, the ongoing work was shut down for safety reasons. For these reasons, this activity exhibited decreased productivity when there was thunderstorm activity (Fig. 5-23) at the site. These conditions occurred on workday 12 and at the end of the observation period. The sustained productivity was approximately 0.13 work hours/dowel. The cumulative productivity was calculated as 0.145 work hours/dowel (an 11% increase).

Concrete Paving—While the productivity of concrete paving was seemingly unaffected by temperatures, there were certainly cases where it could have been

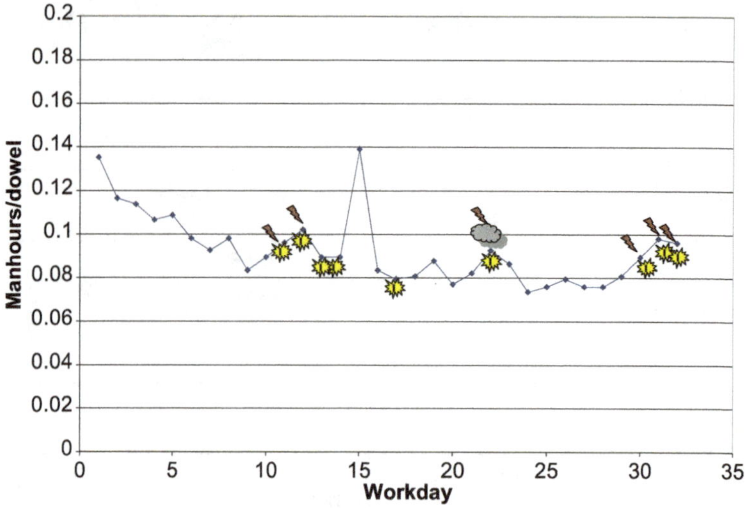

Fig. 5-23. Dowel Bar Retrofit Productivity

affected because high heat, low humidity, and wind can cause the concrete to dry rapidly. In these conditions, productivity is often affected because of the need for a large portion of the crew to take care of the section of pavement recently placed to ensure that it does not dry before it is finished and sealed. Additionally, cold temperatures can prevent the concrete from curing properly. In the case of a concrete slab that loses heat rapidly, there is no practical way to adequately provide heat. These two situations never occurred. But rainfall is another big issue when it comes to weather effects on concrete paving. Even a small amount of rain can ruin concrete pavement if it has not begun to cure. For this reason, concrete paving was seldom scheduled on days when there was a chance of rain. In the event that rain appeared imminent, the operation was concluded quickly and measures such as covering the slab were taken to protect the concrete from rain damage. The crew for this operation was largely pulled from other activities around the job. For this reason, if concrete paving was canceled because of the threat of bad weather, the crew worked on another task (Principle 1.4). This method had a mitigating effect on any reduction in productivity caused by rain and thunderstorms.

The concrete paving operation studied was affected on five days (Fig. 5-24) when there were thunderstorms or threats thereof, most notably, workdays 11 and 15. In the absence of weather events, the crew was able to pave at a productivity of about 0.080 work hours/m^3 or less. The cumulative productivity was 0.088 work hours/m^3 (a 10% increase).

Asphaltic Paving—Asphaltic paving is not as sensitive to rain as is concrete paving. Because of the heat of the material, some rain can be tolerated before the end product is affected. For this reason, asphaltic paving does not show the decreased productivity on rainy days that is seen in concrete paving. Because of the higher labor intensity

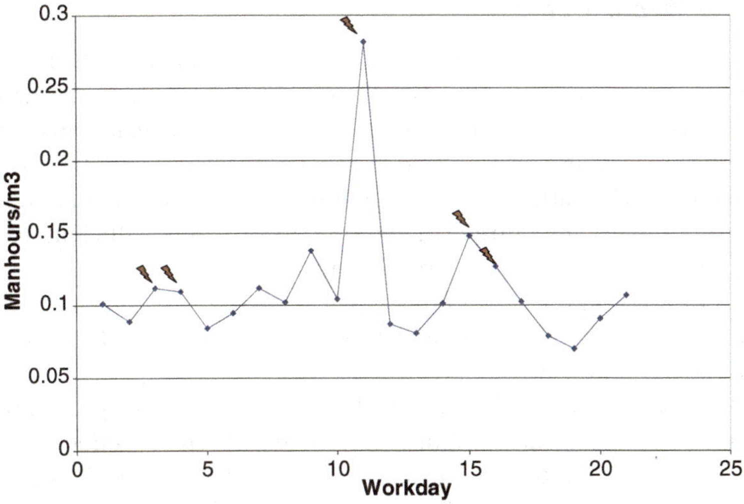

Fig. 5-24. Concrete Paving Productivity

and the heat of the material, asphaltic paving would seem to be somewhat more susceptible to decreased productivity caused by high temperatures. This effect is reduced somewhat by the mechanization of the process, but reduced productivity can still be seen, especially in cases of exceptionally hot weather.

The asphaltic paving operation studied was affected most by high temperatures on workday 9 when the maximum temperature was 89°F (see Fig. 5-25). Otherwise, asphaltic paving was adversely affected by rain and thunderstorms. The sustained, steady-state productivity was about 0.060 work hours/m^3. The cumulative productivity was 0.064 work hours/m^3 (a 6% increase).

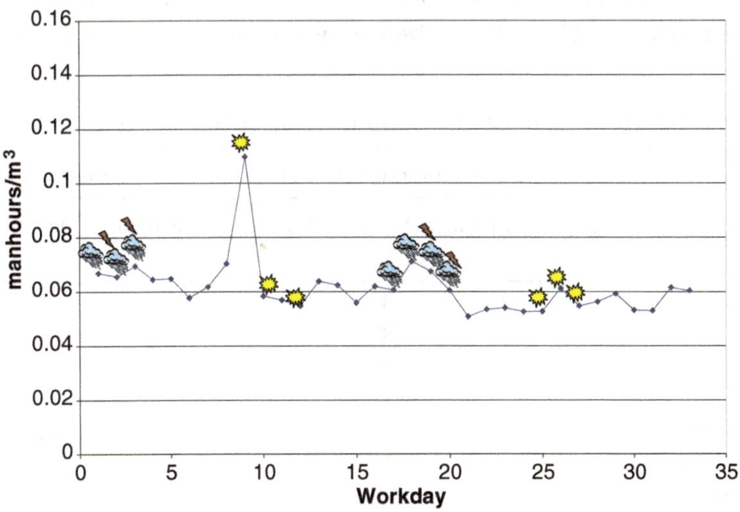

Fig. 5-25. Asphaltic Paving Productivity

Synopsis of Weather Effects

The project selected for analysis was chosen because it was mainly affected by weather and little else. Therefore, the difference between the sustained, steady-state productivity and the cumulative productivity is treated as the approximate effect of weather.

The effect of weather is sensitive to the activity being evaluated. Each operation was affected uniquely. Earthwork showed lasting effects, as long as one to two weeks. For some operations, the threat of a thunderstorm was sufficient to shut the operation down. It was not necessary for it to rain at the site.

Weather events on the case study project showed greater effects than did high temperatures, especially from thunderstorm activity and rain. The effect of high temperatures was minimal.

The effect of weather on labor productivity on the case study project is generally less than was reported in the literature (see Table 2-2), but the effect is close to the value calculated in the estimating example given earlier in this chapter. Table 5-3 summarizes the estimated effects for the activities studied.

Table 5-3 yields some surprising results. The equipment-intensive activities on this case study project showed the greatest effect, and the labor-intensive work showed the least. The equipment-intensive work primarily experienced work stoppages, whereas the labor-intensive work tended to continue through the affected period. Pile driving was affected the most.

5.6 Case Study 2—Interstate 99 Bridge Construction

This case example demonstrates the effect of rain on bridge construction. In this instance, the work continued through the rain events.

The project was the construction of Bridges 28 and 29 as part of Interstate 99 in central Pennsylvania. The activity observed was concrete formwork for abutment

Table 5-3. Estimates of Effects Caused by Weather on Highway Case Study 1

	Equipment Intensive	Labor Intensive
Activity	Estimated Weather Impact %	Estimated Weather Impact %
Sheet piles	35	
H-piles	37	
Tube piles	22	
Earthwork	12	
Dowel bar retrofit		11
Concrete paving	10	
Asphalt paving	6	
Bridge formwork		7

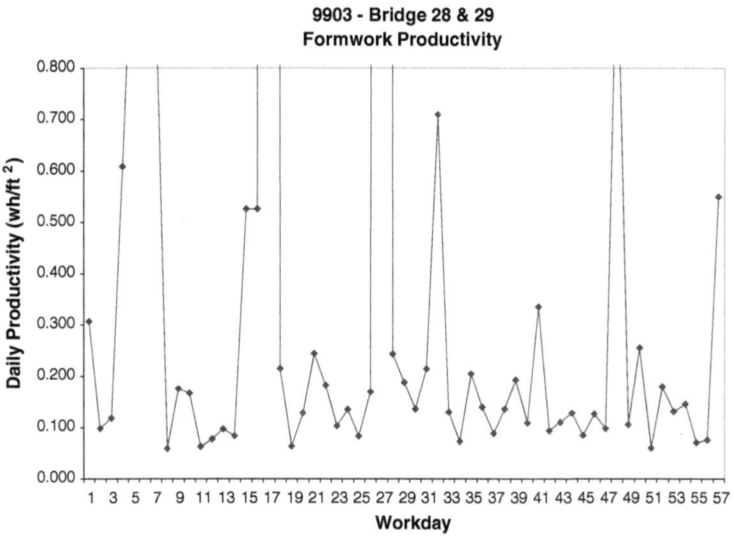

Fig. 5-26. Productivity of Bridge Formwork, Bridges 28 and 29

walls and footers. The crew size varied, but averaged 10–13 nonunion carpenters. The same crew also placed concrete. On days when concrete was placed, additional laborers were usually added.

Fig. 5-26 shows the daily productivity of the formwork operation (Thomas et al. 2002). Rain affected the work on seven workdays (workdays 1, 2, 29, 50, 51, 54, and 57). The total inefficient work hours resulting from rain were more than 290. The inefficiency from rain was estimated as 7.2%, which is less than was reported in the literature. The time of day when the rain occurs is an important factor in determining inefficiency. The work was affected by multiple other factors not related to the weather. The causes of these other effects were not considered.

References

Clapp, M. A. (1966). "The effect of adverse weather conditions on five building sites." Construction Current Paper No. 21, The Building Research Establishment, Watford, U.K, 171–180.

Thomas, H. R., and Ellis, R. D., Jr. (2009). "Fundamental principles of weather mitigation." *Pract. Period. Struct. Des. Constr.*, **14**(1), 29–35.

Thomas, H. R., Horman, M. J., de Souza, U. E. L., and Zavrski, I. (2002). "Benchmarking of labor-intensive construction activities: Lean construction and fundamental principles of workforce management." Rep. 276, International Council for Research and Innovation in Building and Construction (CIB), Rotterdam, Netherlands, 156.

CHAPTER 6

Fundamental Principles of Site Material Management

The discipline of construction project management is often ill defined, particularly when it comes to site operations. There is a considerable body of knowledge about the things that go wrong on the site and the consequences of ineffective decision making, but there is limited information about what procedures and steps a contractor should follow to avoid undesirable cost overruns and time delays. This is particularly true of site material management practices (Thomas et al. 1989, 1999; Thomas and Sanvido 2000). Considering that the amount of materials delivered to the site can be significant, it is important that project engineers develop good material management habits.

The full scope of material management includes quantity takeoffs, submittals, ordering, expediting, tracking, delivery schedules, site storage, handling, waste removal, and other activities. This chapter concentrates on practices from material deliveries to the site forward. Perhaps one reason that current site practices are so ad hoc is that there is nothing written to guide contractors through this process. It is more glamorous to write about large computerized ordering and tracking systems than site housekeeping. The literature focuses mainly on large computerized systems and bar codes (Stukhart and Cook 1990). Thus, earlier writings are of little benefit to many contractors because large, computerized tracking systems are simply not relevant to a sizable sector of the construction industry. This chapter guides project engineers and project managers through the site material management process.

The most obvious risk of ineffective material management is double-handling of materials. But congestion and poor housekeeping can also be problematic. Damaged materials, poor-quality workmanship, and theft can also be troublesome. Another risk often observed by the authors is running out of materials. Each of these deficiencies can seriously degrade labor productivity and increase job costs.

6.1 Goal of Effective Material Management

The goal of effective site material management is to move materials from the point of delivery at the site to the point of installation in the most efficient manner possible.

To achieve this goal, a fundamental practice is to avoid double- and triple-handling of materials. Double- and triple-handling of materials can be readily avoided by eliminating steps in the delivery process. Some double- and triple-handling cannot be avoided, though.

6.2 The Material Journey

Fig. 6-1 conceptually shows the various stages of site material delivery. The material is first delivered from the vendor and then off-loaded at the site. These are shown as points 1 and 2. The material in this scenario may next be stored at the site (point 3). At some time, the material is moved to a lifting device (e.g., a hoist, elevator, or crane), which is shown as point 4. The material is then lifted vertically to what is referred to as a temporary staging area (point 5). Finally, it is moved to a distribution point (point 6), which is located several feet (it is hoped) from where the workers install the material, called the work face (point 7).

In total, Fig. 6-1 shows seven key points or stations and six legs of the journey. Each leg goes from one point to another. The objective of management is to move the materials from point 2 to point 7 in as efficient a manner as possible so that there is no delay or idle time at point 7. To do so, each point must be controlled, and important decisions (strategies) must be made about each leg of the journey.

Procurement

Movement of materials from point 1 to point 2 is largely a procurement function and is sometimes done at the home office. It requires a good site plan, possibly timed

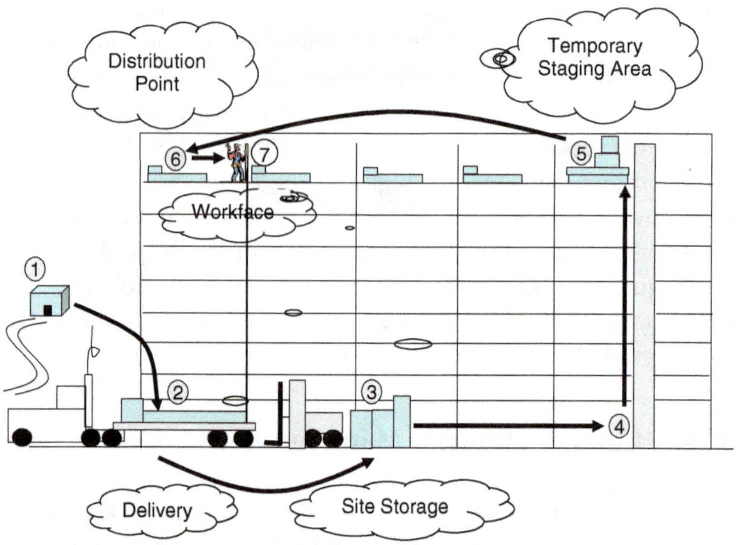

Fig. 6-1. Conceptual View of the Material Journey

deliveries, and good vendor relations and coordination. These latter functions are predominantly site functions.

Delivery Practices

Three practices for delivering materials to the work face (point 7) deserve further discussion. The three practices are (1) delivering materials in large lot sizes or all at once and stockpiling them at the site, (2) the practice of erecting materials directly from the delivery truck, and (3) preloading. Several of the important issues that are germane to these practices are surge piles, stockpiles, off-site staging areas, and vendor coordination. Managers may need to use one practice for one kind of material and a different practice for another. The practice of preloading materials may dictate the strategy used. The overall strategy needs to encompass delivery from point 2 to the work face (point 7). How the materials are loaded on the delivery truck at point 1 may also be important. It is possible through effective planning to eliminate some points, which is a desirable plan.

Deliver Large Lot Sizes—A common practice is to deliver large volumes of the materials and stockpile them on site before beginning the installation. This approach is a least preferred approach. Several disadvantages include the need for storage space (point 3) and the requirement for forward purchases. A third disadvantage relates to the actual delivery process itself. If deliveries are made early, before the installation crew is fully mobilized, there is limited crew idle time during the unloading process. If deliveries are made after the installation crew is fully mobilized, some of the crew will be idle, watching the other part of the crew unload the delivery truck because unloading generally requires only part of the crew. Some of the disadvantages with this strategy may be negated by the effective use of surge piles and off-site staging areas. On the positive side, all the materials are on hand, so the crew should not run out of materials during installation. Material shakeout can also be done at the site. The delivery of structural steel using this strategy is shown in Fig. 6-2. Here, the available storage space is limited by the reach of the crane, so the presence of a large, spacious site may be deceiving. The storage area must be well organized. On the project shown in Fig. 6-2, structural steel is shown stockpiled at the site. An off-site staging area was used, so the steel was delivered to the site about every third or fourth day. When large lot sizes are planned, it is important that the existing stockpile not get too depleted as crews tend to slow their pace of work as the existing stockpile gets smaller and smaller. If deliveries are made throughout the duration of the project, when the installation crew is fully mobilized, care must be exercised to not have a part of the crew watching the unloading process, as the full crew is often not needed to unload a delivery truck. Thus one significant disadvantage of relying on site stockpiling of multiple deliveries is that it may not be possible to fully engage the crew. On most days when materials are delivered, little or no production work is done. It is unbelievable how many times crews actually run out of materials. Also, convenient access must be maintained. It follows that material delivery practices are one of the leading causes of idle time and labor inefficiencies.

Fig. 6-2. Illustration of Storage Requirements When All or Most Materials Are Delivered at Once

Fig. 6-3. Preloading of Drywall

Preload—A favored practice is to preload materials. Fig. 6-3 shows drywall that has been preloaded (point 6). Preloading is the practice of storing just the right amount of material at the distribution point (point 6) before the beginning of installation. Preloading materials directly from the delivery truck can eliminate points 3–5 in Fig. 6-1. The advantage of preloading is that when the crew shows up, all their materials are readily available, no more, no less. All the crew has to do is begin work. Do not forget incidental items, like screws and nails. The disadvantage of preloading is that it takes careful planning. However, the benefits to labor efficiency can be significant.

Preloading merits special discussion. Preloading usually applies, but is not limited to, materials used by the service and finish trades. Preloading should not be confused with the practice of using constructed spaces as storage areas. To preload, material fabrication and delivery schedules may need to be accelerated.

Several figures help to illustrate the concept of preloading. Fig. 6-4 shows an effective preloading practice. Only the right amount of material needed is stored on the floor, the area is clear of trash and debris, and scaffolding is readily available. When the crew moves into the area, all they have to do is work. Fig. 6-5 is an example of good practice of preloading duct.

Fig. 6-4. Effective Preloading Practice

Fig. 6-5. Preloading of Duct

Fig. 6-6 shows an ineffective application of preloading. While the pallets of floor tiles have been distributed throughout the floor where needed, it has been done prematurely so as to impede the work of the duct crew, framing crew, and perhaps others. The pallets have been moved in Fig. 6-7, leading to costly double-handling.

Figs. 6-8, 6-9, and 6-10 should not be confused with preloading because these constructed areas are being used for semipermanent storage.

Preloading of all needed materials at once may not be practical. Fig. 6-4 shows the preloading of materials for multiple crafts. Fig. 6-5 shows the piecemeal

Fig. 6-6. Ineffective Application of Preloading

Fig. 6-7. Ineffective Application of Preloading

Fig. 6-8. Use of Constructed Areas for Storage

Fig. 6-9. Storage of Materials in the Building Interior, SCMB

preloading of duct only. Presumably, preloading for other crafts will be done at a later time. Regardless of the approach taken, preloading is a highly beneficial practice. Preloading on a second shift may be considered. If piecemeal preloading is done, one must still solve the problem of how to deliver the materials from the temporary staging area (point 5) to the distribution location (point 6) close to the work face (point 7). It is undesirable for crafts to travel more than a short distance to procure materials. Movement beyond point 5 is efficiently done with dollies. Manual transport should be the option of last resort.

Fig. 6-10. Use of the Third Floor as a Semipermanent Storage Area

Erect Directly from the Delivery Truck—Another efficient practice is to erect materials into place directly from the delivery truck (Thomas et al. 1999). This practice can eliminate points 3, 4, 5, and 6 in Fig. 6-1. The disadvantage is that deliveries are made daily or at least more frequently. However, a significant advantage is that no on-site storage space is required. Obviously, this practice must be well planned to ensure that the right items are on the truck and that they are loaded in the order needed. If fabricated items are being installed, close coordination with the vendor is essential. Another advantage of this method is that there is little or no crew downtime spent watching others unload the delivery truck. Shakeout must be done off site. Fig. 6-11 shows erection directly from the delivery truck of a precast column.

Fig. 6-11. Erection of a Precast Column Directly from the Delivery Truck

An efficient delivery strategy combines this approach and preloading, thereby eliminating points 3–5, and possibly point 6.

If erection from the delivery truck is adopted, then it may be necessary to develop a timed delivery schedule. Case Study 3 at the end of this chapter illustrates the effective use of a timed delivery schedule.

6.3 Delivery of Materials to the Work Face

Once materials are delivered to the site, the delivery strategy is not complete. The materials must be delivered to the work face (point 7). The engineer must decide how each leg in Fig. 6-1 will be accomplished in an efficient manner. Once decking or precast slabs are installed, a crane will be of limited use. For some smaller facilities, a forklift or high lift may be usable. For taller buildings, an elevator may be the preferred option to deliver materials to point 5. The service elevator may be usable for some smaller items. Don't forget the movement to the distribution points (point 6).

6.4 Division of the Site

Relative to the management of materials, a construction site should be divided into three major areas: a semipermanent storage area, a staging area, and a work face area.

Semipermanent Storage Area—These are areas, sometimes called laydown areas, where materials are stored (stockpiled) before being used in the project (see Fig. 6-2). This is represented by point 3 in Fig. 6-1.

Staging Area—This area is adjacent to the exterior of the facility. It is in this area that materials are lifted into the facility. Materials that are off-loaded directly into the facility from a delivery truck also use this area.

Work Face Area—This is the area inside the facility where the work of the craftspeople takes place (point 7).

Each of these areas has a different use and must be managed differently; also, different principles apply in each area.

6.5 Fundamental Principles

Fundamental principles for material management are summarized in Table 6-1. They are organized according to the three areas of the site related to material management. It should be obvious that compatibility of principles with site plans (see Chapter 4) is essential.

Table 6-1. Principles of Site Material Management

Number	Principle
	Semipermanent (Outside) Storage Area
1.1	Do not store materials close to the building.
1.2	Locate parking areas, toolsheds, trailers, and spoil piles as far from the building as practical.
1.3	Mark stored materials so that they can be readily distinguished from similar materials.
1.4	Materials should be stored to permit easy access and retrieval.
1.5	Materials should be stored on timbers or pallets to prevent damage from mud and water.
1.6	Make effective use of surge piles (stockpiles) to ensure that work (components) are always available for the crew.
1.7	Avoid multiple staging areas at the site because this can lead to double-handling of materials and inefficiencies when unloading materials at the site.
1.8	Whenever possible, especially if storage space is limited, consider final staging of large materials at an off-site location.
	Staging Area
2.1	Reserve areas next to the building for material deliveries or materials being moved to the work face areas.
2.2	Backfill around the building as soon as practical to permit the area to be used as a staging area.
	Work Face (Interior) Storage Area
3.1	The amount of material stored inside should be kept to a minimum.
3.2	Preassemble components into larger components or subassemblies.
3.3	Preload materials and distribute throughout the work face area so as not to interfere with other work.
3.4	Integrate the sequence of work with the storage plan so that interior space can be used without interfering with other work.
3.5	Ancillary tasks like unpacking, cutting, reshaping, and preassembly should be done away from the work face when practical.
3.6	Maintain good housekeeping.
3.7	Arrange for removal of waste from the building on a continual basis.
	Vendor Relations and Deliveries
4.1	Whenever possible, erect deliveries directly from the delivery trucks.
4.2	Ensure that deliveries are properly sequenced to be consistent with the work plan.
4.3	Make sure that the delivery rate from vendors is compatible with the installation rate of the field crew.
	General
5.1	Avoid the use of earthen ramps into below-grade areas.
5.2	Use elevators.
5.3	Have a good site plan for the key phases of the work.

Semipermanent Storage

Semipermanent storage areas are where materials are stored for an extended period of time, say several weeks or more. These areas are usually thought of as being outdoors, but this is not necessarily always the case, as shown in Fig. 6-10. On this building project, the third floor was used as a semipermanent storage space. Importantly, materials should not be stored next to the building (Principle 1.1), because these materials can impede normal access requirements, including access for cladding (see Fig. 3-25), scaffolding, material deliveries, and access into the building (Figs. 6-12 and 6-13). One can see that in Fig. 3-3, parking areas, spoil piles, trailers, and toolsheds all occupy valuable space on the State College Municipal Building (SCMB) project. Clearly some judgment must be applied in the location of trailers, parking areas, and toolsheds, but generally these should be located as far

Fig. 6-12. Congested Staging Area, State College, PA

Fig. 6-13. Poor Material Storage Practices, Rider Building, State College, PA

away from the facility as is practical (Principle 1.2). Conversely, other options may be possible for spoil piles and parking areas. The real estate in a restricted site is much too valuable for these uses.

Materials in the storage areas should be clearly marked (Principle 1.3) so that they can be readily distinguished from other similar materials. For example, a 21-ft piece of steel reinforcement should be easily distinguishable from a 22-ft piece of the same configuration. Fig. 6-13 shows an example of poor storage practices. The labor productivity of the steel erection crew on this project was much worse than on comparable projects, and much of the cause can be attributed to deficient material storage practices (Thomas et al. 1989).

Materials should be stored in a manner that will allow easy access and retrieval by lifting and transporting equipment (Principle 1.4). This strategy is clearly difficult on the project shown in Fig. 6-13. The materials were not segregated, and access and retrieval were difficult at best. Fig. 3-11 shows a mixture of good and marginal storage practices. Some reinforcement is marked and stored on timbers; some is not. Principle 1.5, store materials on timbers and pallets, has been inconsistently applied in Fig. 3-11, and easy retrieval is not always possible. Fig. 6-14 is another example of poor storage practices. Materials should also be stored on timbers or pallets to prevent damage from mud and water (Principle 1.5). Figs. 5-4 and 6-15 show two projects where it was almost impossible to protect materials from mud and water. The performance on both projects relative to time and cost was poor.

In certain instances, contractors can make effective use of surge piles (Principle 1.6) to ensure that the crew does not run out of materials. This method does not automatically mean delivery of all the materials to the site at once. Surge piles may be at an off-site location. Case Study 3 at the end of this chapter describes the

Fig. 6-14. Inaccessibility of Stored Materials, Walker Building

Fig. 6-15. Muddy Site and Storage Conditions, Walker Building

effective use of surge piles, whereas Case Study 4 describes the ineffective use of a surge pile.

In most cases, multiple staging areas should be avoided (Principle 1.7), and whenever practical consider final staging at an off-site location for selected materials (Principle 1.8) and erection directly from the truck. Using multiple staging areas leads to more material handling. More material handling can result in increased theft, damage, and misplacement or loss of materials.

Staging Areas

Areas immediately adjacent to a building perimeter are needed for multiple functions. Considerable work on the exterior cladding and windows is performed there, as are crane locations and concrete discharge points. Also, open areas are needed to ensure easy access for deliveries of equipment and materials and for waste removal. Materials stored in the staging area are more likely to require double-handling leading to Principle 2.1, which is to store materials elsewhere.

Stockpiles of excavated material should be located away from the building, and backfilling around the building should be done as soon as practical (Principle 2.2). On a small site, consideration should be given to stockpiling excavated material off site, as a stockpile of material will occupy valuable space. Late backfilling around the perimeter of a building can be observed in Fig. 6-16. This method violates Principle 2.2 and can inhibit access to the building.

Work Face (Interior) Storage Practices

Obviously, some materials must be stored inside a building or facility. However, the amount stored here should be kept to a minimum (Principle 3.1). Only one or two

Fig. 6-16. Late Backfilling, O'Leary, Lewisburg, PA

day's supply should be stored inside unless the areas are no longer needed or unless preloading is practiced. The consequences of using an active work area for storage are clearly seen in Fig. 6-9. Fig. 6-17 shows another example of inside storage that can impede progress. This project underwent considerable distress.

Whenever possible, preassemble smaller components into larger components or subassemblies (Principle 3.2). This concept is illustrated in Figs. 2-2 and 6-18.

Preload materials into the facility so that materials are readily accessible to the crafts (Principle 3.3). Avoid carrying materials one piece at a time (Fig. 6-19) up a stairwell. This is a strategy of last resort and is highly inefficient and undesirable. Distribute the materials along the work face area, but not so as to interfere with other work (Fig. 6-4). If improperly distributed, double-handling will be the result.

The sequence of the work should be integrated with the storage plan (Principle 3.4). Riley has written at length about the need to do so (Riley and Sanvido 1995, 1997). On some projects, the basement area can be used for inside storage. Do not store materials in areas that will be needed later.

Fig. 6-17. Library Storage, West Point School Renovation, Pittsburgh, PA

Fig. 6-18. Duct Subassembly

If care is not taken, the work face area can become cluttered with excess material and waste. Tasks such as unpacking and cutting and reshaping materials should be limited in work face areas (Principle 3.5). Assign these tasks to other locations, such as the staging or storage areas. Fig. 6-20 shows an example of an untidy work face area. This unpacking work could easily be done elsewhere. If done in the work face area, cleanup becomes important.

Housekeeping is an important function that should be given close attention (Principles 3.6 and 3.7). Poor housekeeping inhibits good productivity and safety. Figs. 6-20 to 6-22 show examples of poor housekeeping. On each of these projects,

Fig. 6-19. Carrying Duct One Piece at a Time

Fig. 6-20. Poor Housekeeping in a Work Face Area

the performance of the particular activity being studied was impaired by poor housekeeping. Poor housekeeping is a commonly observed problem. Fig. 4-4 shows a messy site where it will be difficult to find and retrieve materials and where movement of crafts and equipment around the site will not be easy.

Vendor Relations and Deliveries

Several technical articles have been written about material deliveries and vendor relations (Thomas and Sanvido 2000). In one article related to structural steel deliveries, it was concluded that the most productive method of delivery was to erect

Fig. 6-21. Poor Housekeeping at a Distribution Point, Hampton Inn, State College, PA

Fig. 6-22. Poor Housekeeping at a Staging Area, Sassafras Project, Selinsgrove, PA

Fig. 6-23. Erecting Precast Element Directly from the Truck (Right Piece, Right Time)

the steel directly from the truck as it was delivered to the site (Thomas et al. 1999). Fig. 6-23 shows the erection of a precast element directly from the delivery truck (Principle 4.1). This method eliminates the need for storage areas on site, but to use this method, control of the staging area must be maintained and there must be close coordination with the vender (see Case Study 1).

The sequence of materials delivered from a fabricator or vendor must be compatible with the erection or installation sequence (Principle 4.2). Otherwise, there will be an accumulation of unused pieces at the site. Consider storing unused and defective pieces at an off-site location.

The delivery rate from vendors must be compatible with the installation rate in the field (Principle 4.3). All too often, these two functions are not well coordinated. Fig. 6-24 shows the daily productivity of a duct installation crew on the SCMB project. Because the deliveries from the vendor were not sequenced properly, the crew had to move to alternate locations, often where other crews were working (Han and Thomas 2002). This situation occurred on workday 15 and workdays 20–22. The negative effect is clearly evident. Thomas and Sanvido show the results of three case studies where deliveries from vendors were not well coordinated (2000). The labor overruns ranged from 17% to 57%.

General Principles

Several principles are important as they relate to transporting materials from the storage or staging areas to the work face areas. The first principle relates to projects with below-grade excavations such as basements. The use of earthen ramps should be avoided because they often impede the completion of basement walls, which may in turn affect the entire project schedule (Principle 5.1). The interference between wall

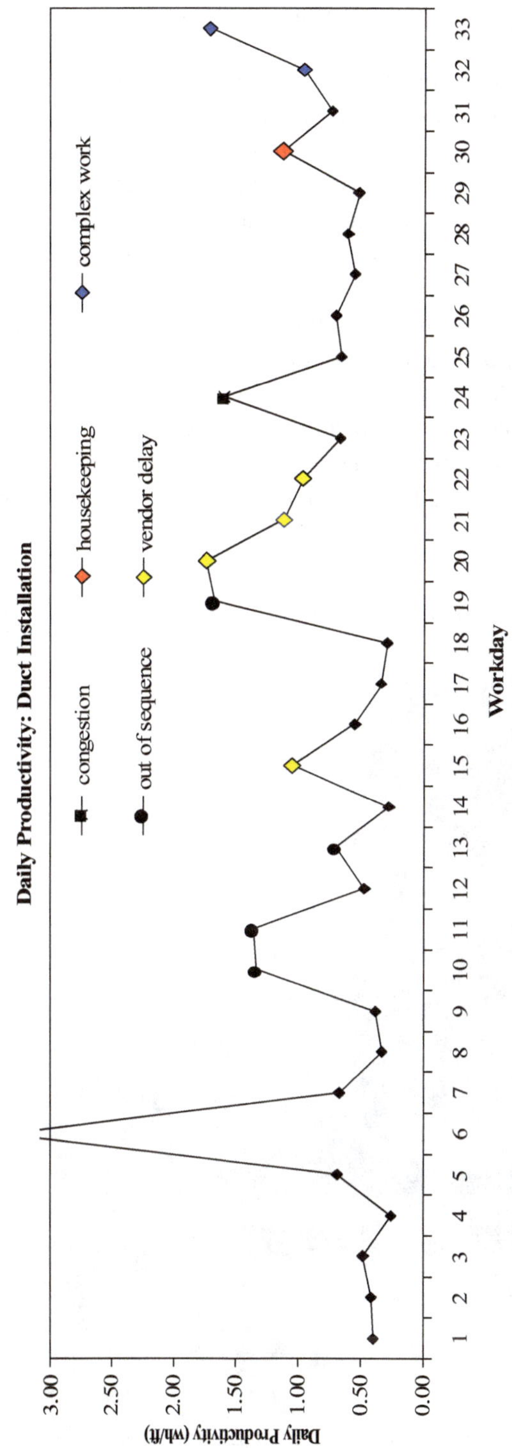

Fig. 6-24. Daily Productivity of Duct Installation, SCMB

construction and the ramp has been observed on several projects. The completion of basement walls was slowed on the projects in Figs. 3-11 and 6-15. An economic analysis may be justified. But given the cost of schedule delays and labor costs, it is hard to envision an alternate strategy that is not cost-effective.

Another principle is that contractors should make use of service elevators and hoists (Principle 5.2). On some projects, cranes and forklifts can be used. Service elevators should be installed as early as practical. On some projects, it has been observed that materials were manually moved from one floor to another because forklifts would not reach the upper floors. Notice that in Fig. 6-25, the staging area must be clear and there is a limit to the height that can be reached. This approach is not always adequate if work activity occupies the staging area and the right equipment is not available when needed. Tight control of the staging area is required. Moving materials into the facility after regular working hours may be discouraged because of overtime pay. However, lifting capacity availability may be restricted during regular working hours, making overtime work the only alternative.

Fig. 6-25. Use of a High Lift to Move Materials into the Building, Paterno Library, Pennsylvania State University, State College, PA

A contractor can ill afford a crew being idle for several hours the next morning waiting for materials to be delivered, yet this sometimes happens.

It is important to have good site plans (Principle 5.3) and enforce the plans. It is easy for infringements to occur. Even seemingly small infringements have the potential to halt key activities and delay progress.

6.6 Developing Material Distribution Plans

The project engineer should develop material distribution plans for each major material that will be delivered to the site. Plans should be developed by the general contractor in conjunction with applicable subcontractors and specialty contractors. An efficient material distribution plan that is rigorously followed will go a long way toward establishing a team environment.

The goal of developing a material distribution plan is to determine how to move materials from their delivery point to the work face efficiently (in this instance, from point 2 to point 7 in Fig. 6-1) so that workers do not have to search for and retrieve the materials they need, thus minimizing crew idle time. Material distribution plans help create a team environment and lead to greater craft efficiency, lower costs, reduced idle time, and much more. Table 6-2 gives a simple eight-step procedure for developing a material distribution plan. The steps are briefly described and are illustrated using the services work on the SCMB.

1. Select Mode of Material Movement (Points 2–5)

This step involves defining how materials will be moved from point 2 to point 5 in Fig. 6-1. It may involve several steps as there may be one mode from point 3 to point 4 and another to point 5. For instance, one may use a forklift to move materials from a storage pile (point 3) to within reach of a crane (point 4). It may be lifted then via crane to a temporary staging area (point 5). If materials are to be erected directly

Table 6-2. Procedure for Developing a Material Distribution Plan

No.	Procedure
1.	Select mode of material movement.
2.	Determine location of temporary staging areas.
3.	Determine location of distribution points.
4.	Determine distribution point quantities.
5.	Develop a work (sequence) plan.
6.	Develop a space allocation layout.
7.	Distribute materials and incidentals.
8.	Communicate and enforce the plan.

from the delivery truck, then there may be only one step. If a high lift is being used, it may be a one- or two-step process. Elevators may also be used. One option that should be avoided is to transport components one at a time. This method is highly inefficient and can be unsafe. Manual transport is a huge red flag that the project is probably not being managed as well as it should. A key point is to avoid double- and triple-handling of materials.

The next question is a matter of scheduling. It may be necessary to preload all materials for multiple crafts early in the construction schedule (Fig. 6-4), or it may be done for a single craft (Fig. 6-5) just before the crafts arrive. In the latter approach, deliveries to point 6 are made frequently, perhaps daily. The mode of movement selected in step 1 is one factor, but the key goal is that the materials must be at their assigned location before the workers arrive for work.

For the SCMB, it was decided in Chapter 3 that the service materials (e.g., sprinkler piping, conduit, framing, and drywall) would be preloaded in conjunction with the steel erection and immediately after the erection of the precast floor planks. A crane was used and erection of structural steel was directly from the delivery truck, so there was no room allocated for on-site storage. Each subcontractor needed to have their materials delivered to the site as steel was being erected.

2. Determine the Location of Temporary Staging Areas (Point 5)

Temporary staging areas are seen in Fig. 6-1 as point 5, and this step involves pinpointing the location of point 5. Temporary staging areas are usually thought of as holding areas where materials are temporarily stockpiled for a short period of time. There can be one or more locations in a given area or on a particular floor, depending on the mode of delivery to point 5.

The plan for the SCMB was to erect directly from the delivery truck using a crane; all preloaded materials for multiple crafts were delivered at once (Fig. 6-4). Fig. 6-26 shows that the floor area was divided into five zones, one in the corridor and four others in the remaining areas. There are six planned temporary staging areas: two in the corridor and one for each the other four zones. Framing, duct, piping, and electrical materials were preloaded initially, as soon as the slabs were complete. Erection directly from the delivery truck was also practiced. The remaining service materials, such as insulation, drywall, and acoustic tiles, were preloaded later.

3. Determine the Location of Distribution Points (Point 6)

This step involves the location of point 6 in Fig. 6-1 and the determination of how materials will be moved from point 5 to point 6. It is not desirable for crews to travel to temporary staging areas (point 5) to retrieve materials as these areas may be far removed from the work face (point 7). Therefore, each zone is subdivided, and distribution locations are identified (point 6). For a commercial building or condominium, subdivision can be done according to room. The distribution points may be

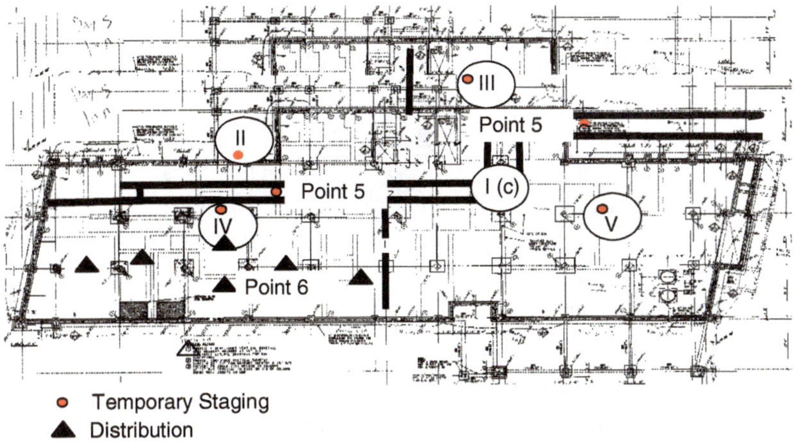

Fig. 6-26. Temporary Staging Areas and Distribution Points, Second floor of the SCMB

along the exterior wall (Fig. 6-4) or in the center of the room (Fig. 6-3). Thus, for an area or floor, there may be only a few temporary staging areas (point 5), but there can be multiple distribution points (point 6). The materials at the temporary staging area are subdivided among the various distribution points. In Fig. 6-26, six distribution points are shown for zone IV. Careful planning can possibly eliminate point 5.

4. Determine Distribution Point Quantities

Because the materials need to be distributed from the temporary staging areas to the distribution points, it is necessary to determine the quantities needed at each distribution point. For example, a room that is 20 × 20 ft has 80 ft of perimeter wall and may require 20 drywall boards. The important point to make is that the amount placed at the distribution points is just the right amount, not too much, not too little. Too much will become clutter and lead to housekeeping issues. Too little will result in idle time when workers run out of materials. Don't forget incidental items, such as screws and nails.

5. Develop a Work (Sequence) Plan

The development of a sequence plan is discussed in Chapters 3 and 10, which show the movement of the crews from area to area. In Fig. 6-26, the order of work proceeds as follows: Ic–II–III–IV–V.

6. Develop a Space Allocation Layout

Materials at distribution points are not randomly stored in any haphazard manner. Materials should be stored so that each craft can retrieve their materials without moving another's. Fig. 6-27 shows a sample space allocation layout where materials

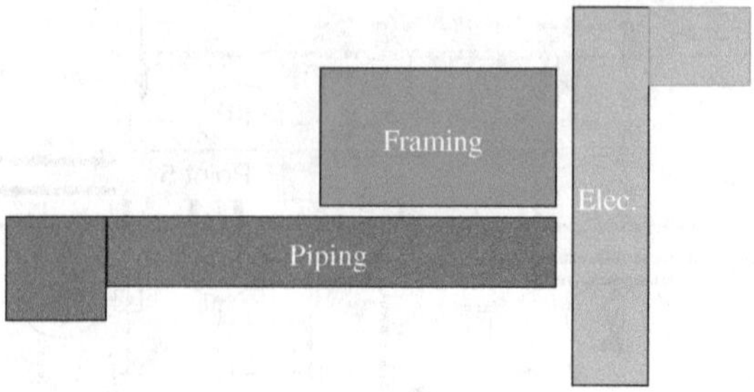

Fig. 6-27. Space Allocation Plan, SCMB

for multiple crafts are distributed at once. A convenient way to communicate the plan is to spray-paint the layout on the concrete slab.

7. Distribute Materials and Incidentals

The materials are typically moved from the temporary staging area to the distribution points manually, e.g., with the aid of dollies.

8. Communicate and Enforce the Plan

The tendency on a construction site is that any open space can be used by anyone for any legitimate purpose. If this attitude is allowed to prevail, the work area will become cluttered by materials, tools, and trash. Clutter is detrimental to craft productivity and safety. The prime contractor or construction manager must enforce the plan to avoid congestion.

6.7 Waste Management

Many items are delivered to the site in packaged containers. These items produce varying amounts of trash and rubbish (Fig. 6-20). There must be a plan to remove trash to avoid clutter and perhaps unsafe working conditions. Sometimes there are contractual requirements that subcontractors must keep their work areas clean. These provisions must be rigorously enforced. Enforcement on a continual basis is preferred, not simply at the end of the day or week. The general contractor will have areas to keep clean as well. Removal from the site is an important part of material management. There should be a regular schedule for removal of trash.

6.8 Case Study 1—Benefit–Cost Analysis: The Rider Building and Greenwich Court

Almost all the principles in Table 6-1 have some potential cost implications, even though most are minuscule. A logical question is why more contractors do not apply better material management practices if these practices will save money. The authors have no answer. When the cost of labor is considered, almost any strategy will be cost-effective.

Suppose a contractor can spend $100 to buy new drill bits and new bits will save an estimated 10 work hours. At $35/h (burdened), the benefit–cost ratio can be calculated as 3.5, denoting a wise expenditure.

The contractor receives a visible invoice for $100, but the payroll amount of $350 is largely invisible because it can be argued that the workers would have to be paid anyway. The tendency of many contractors is to spend monies on visible expenditures and to ignore invisible benefits. So, the drill bit purchase may not be made, despite a favorable benefit–cost ratio.

This case study compares the structural steel erection practices on two almost identical commercial projects. The comparison illustrates the value of several principles in Table 6-1 and demonstrates the need to include inefficient labor (invisible) costs in any economic analysis (Thomas et al. 1989).

Rider Building

Project Description—The Rider Building is a five-story commercial office building constructed in downtown State College, Pennsylvania, in 1984. The building consists of a structural steel frame and brick facade. A new, two-story steel frame parking deck is adjacent. The site was small and constrained. The total site area was 38,950 ft^2, and the combined footprint area of the two structures was 23,920 ft^2, or more than 60% of the total site area. Thus, the area available for material storage and construction equipment was limited. The material delivery strategy of the contractor for structural steel was to make three deliveries of steel. The last two deliveries were made during the erection process, and the material was stored on site. Thus, the erection crew was fully mobilized for the second and third deliveries. There was some idle time during these deliveries, and there was limited or no production during days when steel was delivered.

Fundamental Principles—The following principles in Table 6-1 do not apply to this project or were not observed: 1.2, 1.6, 2.2, 3.1, 3.2, 3.3, 3.4, 3.5, 3.7, 4.3, 5.1, and 5.2. The contractor applied Principles 1.3 and 1.7. Principles 1.1, 1.4, 1.5, 1.8, 2.1, 3.6, 4.1, 4.2, and 5.3 were not applied.

Contractor Operations and Practices—The facility was constructed by a local contractor for approximately $5 million in 1985 using a nonunion workforce. The site staff consisted of a single project superintendent. Most material requisitions were made through the home office located approximately 50 miles from the project site. Limited coordination with the fabricator relative to deliveries was evident.

Structural Steel Characteristics—The steel members were mostly in the range of 14 in. × 22 lb/ft to 21 in. × 68 lb/ft. In all, the structure consists of 414 pieces weighing 171.7 tons. Incidental pieces, such as sag rods and gusset plates, are not counted. The connections are AISC type 3 pinned joints with an average of two rows of bolts (a total of four bolts) per connection. Alignment of the structure was done with turnbuckles and guy wires. Final tightening of bolts was done manually with a torque wrench.

The work was done by a four-person crew, consisting of a supervisor and three ironworkers. A crane operator was also present, but his hours are not included in the case study. The steel was initially erected with a 15-ton P&H hydraulic crane. This crane would have had limited reach and would have been moved multiple times. This step probably violates Principle 1.9 in Chapter 3 on planning, but this aspect was not a part of this study. There was some crew idle time associated with each movement of the crane.

The steel was delivered in three separate shipments at different times during the erection process, once before erection began and again after workdays 13 and 33. The crew ran out of steel twice. Coordination with the fabricator was lacking. There was no evidence of a site layout plan. Each time there was a delivery (workdays 14 and 34), the steel was off-loaded in a haphazard manner wherever there was space available. The contractor relied on the "dump truck" method of stockpiling steel. The general condition of the site is seen in Fig. 6-13. As can be seen, the site was in a state of considerable disarray. Movement of the crane would have been difficult.

Contractor Performance—The steel erection was performed by the general contractor. The work lasted 37 workdays and required 1,256 work hours (not counting the work hours of the crane operator or crew supervisor). The daily steel productivity is shown in Fig. 6-28. The productivity is reported as work hours/piece of steel. Therefore, low numbers show better performance. Given the variability in the data, it is readily obvious that this was not a particularly good project.

Fig. 6-28 is revealing about this project. The crew ran out of steel at the end of workdays 13 and 33. After workday 13, there was a five-day delay before steel for the remaining floors and the roof was delivered (the five days after workday 13 are not

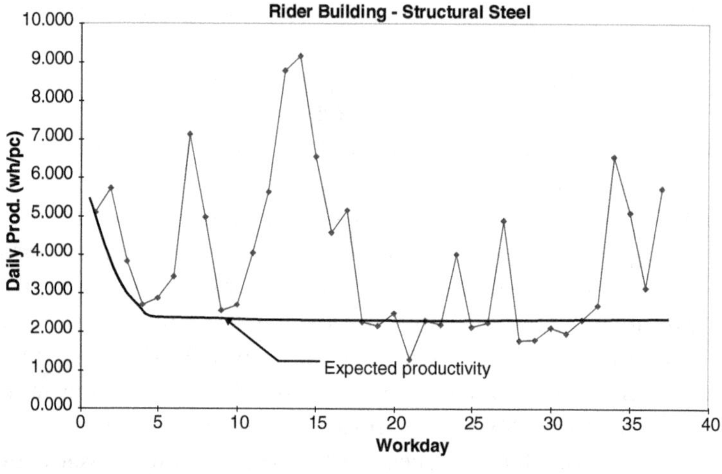

Fig. 6-28. Daily Productivity of Steel Erection, Rider Building

included in this study). Structural steel was delivered on three consecutive days. Because the crew was fully mobilized, there was some crew idle time associated with the delivery. Steel erection commenced after this delivery was made, while the crew was largely struggling to organize the materials (an estimated 64 inefficient work hours). Ten workdays were lost from the schedule. Before exhausting the supply of steel, there were periods of time when the crew slowed its work pace to keep from running out of work. Both times the crew productivity suffered. Whenever steel was delivered, there were periods of about four workdays before the operation returned to its normal productivity level (expected productivity, see Fig. 6-28). During this time, the crew was organizing the steel as it was searching for the next piece of steel to erect. Considerable time was spent moving the steel and other materials in search of the required piece. The figure shows the effect of this practice on labor productivity.

On workday 33, the crew again ran out of steel, and its normal work pattern was again interrupted. Steel for the parking deck was not delivered until 85 workdays after the steel erection on the office structure was complete. While the delivery was only 27 pieces, representing 6.5% of the total, the crew had to remobilize. By this time, the site was congested, making it even more difficult for the crew to locate and organize its materials.

The crew also had difficulty aligning the first two floors, which occurred during workdays 5–9. The situation was corrected, and difficulties with alignment did not occur thereafter.

Overall, the work seemed poorly planned. For instance, on several workdays, few or no pieces of steel were erected and most of the work on these workdays involved tightening, a minor but essential subtask. This may have been "busy" work. Tightening was necessary, but it required at most two ironworkers. So there was likely more idle time or "busy" work. This topic is discussed in Chapter 7.

Greenwich Court

Project Description—The Greenwich Court Building is also a five-story commercial office building constructed in downtown State College, Pennsylvania (Thomas et al. 1989) in 1985. The building consists of a structural steel frame and a brick facade. The site was small and constrained, leaving little area available for material storage and construction equipment. The material delivery strategy of the specialty contractor was to make daily deliveries of steel in the morning and erect the steel directly from the delivery truck. A surge pile of steel was maintained for afternoon work because no deliveries were made after lunch. This stockpile was small and well organized. The contractor coordinated with the fabricator about which pieces of steel were to be placed on which delivery truck. The contractor also told the fabricator which piece to load first so that the piece that was needed first was always on top. The order of deliveries was compatible with the erection sequence.

Fundamental Principles— The following principles in Table 6-1 do not apply to this project or were not observed: 1.2, 1.6, 2.2, 3.1, 3.2, 3.3, 3.4, 3.5, 3.7, 4.3, 5.1, and 5.2. The contractor applied principles 1.1, 1.3, 1.4, 1.5, 1.7, 2.1, 3.6, 4.1, and 4.2. Principles 1.8 and 5.3 was not applied.

Contractor Operations and Practices—The facility was constructed by a local specialty contractor using a nonunion workforce. The site staff consisted of a single project superintendent. The home office was located approximately 30 miles from the project site.

The work was done by a three- to four-person crew, consisting of a supervisor and two to three ironworkers. The crane operator and crew supervisor work hours are not included in the study.

Contractor Performance—The steel erection on Greenwich Court lasted 22 workdays and required 526 work hours. The daily steel productivity is shown in Fig. 6-29. The productivity is reported as work hours/piece of steel. Therefore, low numbers show better performance. Given the variability in the data, it is readily obvious that this was a much better project than the Rider Building. This fact is confirmed in Fig. 6-30, which shows the cumulative productivity of the two projects. Greenwich Court outperformed the Rider Building by a wide margin and also took much less time, 22 workdays compared with 37 workdays for the Rider Building, even though both buildings consisted of approximately the same amount of work.

Fig. 6-29 is revealing about the Greenwich Court project. The crew ran out of bolts, and the poor performance on workday 18 resulted because the crew had to return to many joints to add and tighten more bolts. Six inches of snow fell on workday 12, and little progress was made that day.

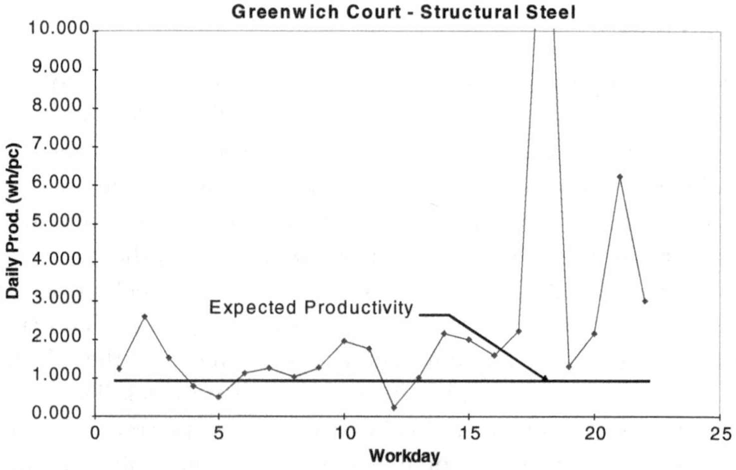

Fig. 6-29. Daily Productivity of Steel Erection, Greenwich Court

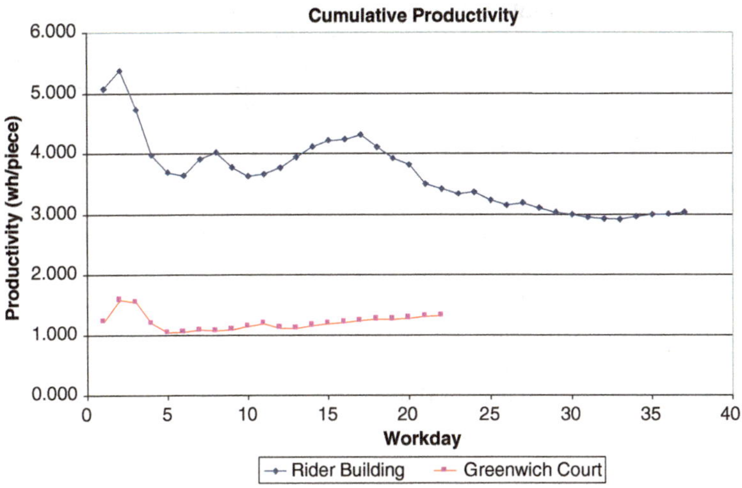

Fig. 6-30. Comparison of Cumulative Productivity, Rider Building and Greenwich Court

Overall, the work at Greenwich Court was well planned. Storage requirements were minimal. The crew knew their work assignments, and there were no difficulties in alignment. Alignment of the first two floors took only one workday (workday 7).

Material-Related Impacts—Calculating the loss of labor efficiency begins by defining the expected daily productivity as the estimated sustained productivity that could be attained on undisrupted days. This is shown in Fig. 6-29. Material-related inefficiencies were computed by comparing the actual daily productivity to the expected productivity values. The inefficiencies for running out of bolts on Greenwich Court were computed as 8.4 work hours. At a burdened payroll rate of $35/work hour, this equates to less than $300.

Benefit–Cost Analysis

Summary statistics for the two case study projects are given in Table 6-3. The two projects are generally comparable. The cost of ineffective material management is measured in terms of direct labor, equipment costs, and the indirect costs associated with the extended schedule on the Rider Building. Because the structural steel erection for both office buildings was on the critical path and the parking deck (for the Rider Building) was not, it is assumed that the inefficiencies in material management resulted in a 10-day delay in completing the Rider Building project. There was no schedule extension on the Greenwich Court project. The wage rates and equipment rental rates used to calculate the benefit–cost ratio were verified by a knowledgeable contractor. The jobsite and home office overhead rates and some other rates are assumed. The inefficient work hours caused by the material management on the Rider Building were calculated as 200.6. (This figure does not count any inefficient

Table 6-3. Project Summary Statistics

Project Attributes	Rider Building	Greenwich Court
Location	Downtown State College, PA	Downtown State College, PA
Type of structural frame	Structural Steel	Structural Steel
Number of stories	5	5
Building height (ft)	62	61
Footprint area (ft^2)	23,920	17,640
Site area (ft^2)	38,950	35,000
Building-to-site ratio	0.61	0.50
Number of pieces	414	399
Number of tons	171.7	179.7
Responsible entity	General Contractor	Specialty Contractor
Number of workdays required	37	33
Crew size	4	3–4
Labor force	Nonunion	Nonunion
Work hours required	1,256	526
Cumulative productivity (work hours/piece)	3.03	1.32
Construction time frame	Sept.–Feb.	Jan.–Feb.

work hours on the days immediately after the first and second steel deliveries when no pieces of steel were erected.) On Greenwich Court, the inefficient work hours were calculated to be 8.4. The following differential cost impact is calculated.

Contractor Cost	
Craft labor (ironworkers): (200 × (6 – 8.4) × \$35/h)	\$6,727
Operating engineer (10 days × 8 h/day × \$35/h)	\$2,800
Equipment rental (\$1,000/day × 10 days)	\$10,000
Total Direct Cost	\$19,527
Jobsite and home office overhead (10 days @ \$550/day)	\$5,500
Total direct and indirect cost for doing the work the Rider Building way	\$25,027

There are added expenses related to fabricator coordination. These include the additional staff hours needed to sequence steel deliveries and the extended wait time for delivery trucks as steel is off-loaded and erected directly in place. Using 25 truck trips to deliver all the steel on the Greenwich project, the following expenses are estimated.

Staff hours to sequence and expedite deliveries	\$200
Truck delivery charges (8 h of superintendent time @ \$75/h)	\$600
Demurrage: (25 trailers @ \$100/day for 25 days)	\$2,500
Total added expenses for the Greenwich Court way	\$3,300

Using these cost figures, the benefit–cost ratio is computed as 25,027/3,300 = 7.58. Thus, for every dollar spent in planning to erect directly from the delivery truck, the contractor can potentially save in excess of $7.58.

Clearly, when the invisible cost of labor and schedule delay times are considered, the Greenwich Court material management strategy is far more cost-effective compared with the Rider Building strategy.

6.9 Case Study 2—State College Municipal Building: Work Face Practices

This case study project is a three-story municipal office building in downtown State College, Pennsylvania (SCMB). The structural frame of the building is structural steel (aboveground stories) and the facade is masonry. The project also contains a basement with reinforced concrete basement walls. The site area is 70,000 ft^2 (200 × 350 ft), and the footprint area of the facility is approximately 18,750 ft^2 (75 × 250 ft). The project was estimated to cost about $5 million in 2001. The planned construction schedule was about 16 months.

Site Characteristics

The actual site plan for the project is shown in Fig. 3-3. It shows that the site had limited space for material storage and laydown areas and other facilities. Fig. 3-3 also shows how the contractor used the limited space. The use of storage and laydown areas was based on a "first come, first served" philosophy, so use of storage areas was largely unplanned. When the study began, steel reinforcement, structural steel, concrete block, and other materials were already stored on site. There was congestion and interference with trucks and deliveries entering and exiting the site. Drive-through deliveries were not applied. There was no evidence of a site plan.

Description of the Activity

The activity reported herein is the ductwork installation, and the operation covered four floors, from the basement to the third floor. The ductwork activity consisted of three subtasks, that is, hangers, duct erection, and connections. There were four major types of components involved in the work: small and large feeder duct, branch duct, and fire dampers.

Fundamental Principles

The following principles in Table 6-1 do not apply to this project or were not observed: 1.1, 1.2, 1.3, 1.5, 2.1, 2.2, 3.5, 4.3, 5.1, and 5.2. The contractor applied

Principle 1.7, and Principles 1.4, 1.6, 1.8, 3.1, 3.2, 3.3, 3.4, 3.6, 3.7, 4.1, 4.2, and 5.3 were not applied. The main focus of this case study is on work face storage practices and housekeeping.

Construction Methods

The duct crew included one supervisor and two sheet metal workers. During the first two weeks, there was enough clear work space for the crew to preassemble three or four pieces of duct together on the floor and use a lift to install the preassembly (Fig. 2-2). However, as the job progressed, the project became congested with framing, drywall, masonry partitions, and materials stored on each floor (Fig. 6-9). This meant that preassemblies could not be used and the duct was installed one piece at a time instead of as a preassembly. The duct was off-loaded and manually carried to where it was to be stored and then manually carried to the work face.

Fig. 6-24 shows the productivity for the first 33 working days of the ductwork activity (Han and Thomas 2002). After the first few weeks, the work was frequently disrupted. During the first two weeks, the productivity was frequently in the range of 0.30–0.35 work hours/ft when the crew made good use of subassemblies. Thereafter, the productivity on undisrupted days was more in the range of 0.50–0.60 work hours/ft. During this time, conditions at the work face prevented the use of subassemblies. The case study and data from this and other projects form the basis for Principle 3.2 (Table 6-1), that is, preassemble components into subassemblies to maximize productivity (Fig. 2-2). Not being able to use subassemblies on the case study project resulted in an estimated loss of productivity of almost 50%.

Material Management Practices

There was no material management plan for this project. Materials were stored on site in accordance with the layout shown in Fig. 3-3. The site was small, and material site access was at the one location shown. There were trailers, a tool storage shed, excavated spoil, and a parking lot located within the site boundary. Material was transported into the building in two places. Fig. 6-31 shows the material storage area on the north side of the building. This area was used to stockpile excavated soil, accumulate trash, and store miscellaneous materials. The temporary utility pole also impeded the free flow of materials. Because of the limited site storage, duct and other materials were stored inside the building. This aspect was shown in Fig. 6-9. The south side of the building was not available for storage because work on installing underground utilities was late.

Waste materials were removed periodically, but often the building interior was untidy. The plan for waste removal (Principle 3.5) was ad hoc.

The disorderly storage practices on the interior affected the duct installation in several ways (Principle 3.1). Duct could not be erected as a preassembly. Also, the framing, masonry partitions, and piping were done out of sequence. Several of these

Fig. 6-31. Storage Area on North Side, State College Municipal Building

aspects are shown in Fig. 6-9. As can be seen in Fig. 6-24, there were numerous disruptions from congestion, interferences, and out-of-sequence work, which negatively affected the work. Overall, the labor performance of the duct installation was poor.

6.10 Case Study 3—Beaver Avenue Parking Garage: Fabricator Relations

This case study project is the construction of a six-story parking structure in downtown State College, Pennsylvania. There is limited commercial space on the first two floors. The project was built in the spring of 2005 at an estimated cost of $11 million.

The entire structure was precast concrete. A specialty contractor using a Manitowoc 2250 and a 10-person crew of union ironworkers did the precast erection. The structure consisted of beams, columns, spandrels, slabs, panels, and stairs. The columns took approximately 45–60 min each to erect, and the other pieces took about 15–30 min each. A separate team of ironworkers did all the grouting. The erection was divided into five phases, but because of time constraints, only phases 1 and 2 were monitored.

Fundamental Principles

The following principles in Table 6-1 do not apply to this project or were not observed: 1.1, 1.2, 1.4, 1.5, 1.7, 2.1, 2.2, 3.1, 3.2, 3.3, 3.4, 3.5, 3.7, 5.1, and 5.2. The contractor applied Principles 1.3, 1.6, 1.8, 4.1, 4.2, and 4.3. Principles 3.6, 3.7, and 5.3 were not applied. The main focus of this case study is on effective vendor relations, use of a surge pile, development of site plans, and the principle of erection from the delivery truck.

Contractor Operations

Because the construction site was small and constrained, there was little or no room for storing precast pieces at the site. To counteract the small site, the contractor applied principles 4.1, 4.2, and 4.3 from Table 6-1.

There were two types of precast pieces: permit and nonpermit pieces. The distinction was the width. Nonpermit or narrow pieces could be delivered without a transport permit, and wide pieces required a permit. The specialty contractor used a vacant staging area about four miles away to maintain a surge pile (stockpile) of nonpermit pieces. As these nonpermit precast pieces were delivered from the fabricator, the loaded trailer was dropped at this remote staging area.

Permit pieces were delivered daily to the site on a prearranged time schedule (Principle 4.3). Because of permit limitations, permit pieces could not arrive at the site until about 10:30 a.m. The use of the remote surge pile of nonpermit pieces provided workers with work to do at the beginning and end of the shift because the superintendent could call for the nonpermit pieces to be delivered at any time (Principle 1.6). When pieces arrived at the site (of either type), they were erected directly from the delivery truck (Principle 4.1). A nonpermit element is shown being erected in Fig. 6-11 and also in Fig. 3-26. A hypothetical delivery schedule is shown in Fig. 6-32.

There was excellent coordination between the specialty contractor and the fabricator (Principle 4.2). The deliveries were timely, except for one day. The correct pieces were delivered per the erection sequence, and the pieces were oriented on the truck to facilitate erection directly from the truck. Thus, the specialty contractor applied good material management practices. The superintendent had a goal of 14 precast pieces per day (based on an estimated budget of 5.75 work hours/piece), but that goal was often exceeded. The remote surge pile was essential to exceeding this goal.

Time	Type	Mark Number
7:05	Nonpermit	W-SCPG-2204
7:20	Nonpermit	S-SCPG-2203
• • •		
10:30	Permit	T-SCPG-2218
11:00	Nonpermit	C-SCPG-2301

Fig. 6-32. Hypothetical Delivery Schedule, Precast Elements

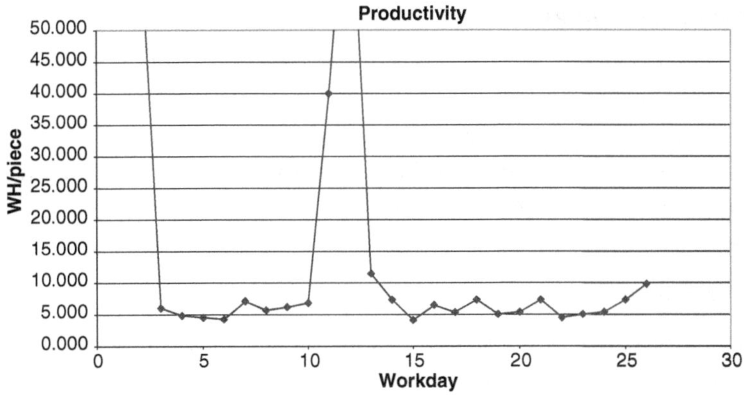

Fig. 6-33. Daily Productivity for Precast Elements, Beaver Avenue Parking Garage

The superintendent had an excellent opportunity to control the daily production because there was good communication and coordination with the fabricator. Most of the deliveries were made on time, and the off-site staging area meant that nonpermit pieces could be delivered as needed, so work for the crew was always available and there was little idle time. The productivity of the crew is shown in Fig. 6-33. Production variability was caused by the time differential needed for different components to be erected: columns took two to four times longer than other components to erect, the stairwells were complex and time consuming, there was an accident (workday 2), design errors affected the work (workdays 7–8), the crane needed to be moved as the work progressed from phase 1 to phase 2 (workdays 11–13), and precast permit pieces were delayed one day (workday 9). Output was low on another five or six workdays. But despite the unfavorable conditions, the goal of 14 pieces per day was exceeded most of the time.

6.11 Case Study 4—Food Science Building: Delivery Strategy

This case study was of the structural steel erection for the four-story Food Science Building classroom and research facility on the Penn State campus. The estimated cost of the facility was $45.1 million. The crew size ranged from seven to 10 ironworkers.

The work was planned in five phases. Two phases were monitored. The overall site was spacious, but to keep steel within the reach of the crane, the on-site storage area was limited. The specialty contractor chose to deliver steel to the site frequently from an off-site staging area, but steel was off-loaded from the trucks and stored on site instead of being erected directly from the truck (Fig. 6-2). Thus, the contractor used multiple staging areas and the final staging area was on site. Because of the U-shape nature of the building, only one crane setup for all five phases was required. Coordination with the fabricator was good, and deliveries to the off-site staging area from the fabricator were timely. The crew never ran out of steel.

Fundamental Principles

The following principles in Table 6-1 do not apply to this project or were not observed: 1.1, 1.2, 1.5, 2.1, 2.2, 3.1, 3.2, 3.3, 3.4, 3.5, 3.6, 3.7, 4.3, 5.1, and 5.2. The contractor applied Principles 1.3, 1.4, and 1.8. Principles 1.6, 1.7, 4.1, 4.2, and 5.3 were not applied. The main focus of this case study is on the use of the surge pile and the material delivery strategy.

Contractor Performance

The labor performance of the steel erection crew on the Food Science Building was not good. The site was spacious, but the storage area for structural steel was limited. The contractor staged the material at two locations: at a remote location about five miles away and at the site. Effective material management practices were not applied. Deliveries were made almost every day, and most crew members had nothing productive to do while steel was being off-loaded. There was much variability in production output. The productivity variability is visually observed in Fig. 6-34. In response to schedule delays, the contractor enlarged the crew, and performance suffered more. The crew performance was variable, especially after the crew size was increased (after workday 10).

It appears that by applying a strategy of staging the steel at the site and increasing the crew size, the contractor seriously affected the labor performance in a negative way.

Presumably, at the off-site staging area, material shakeout was done, as no fabrication errors were observed at the site. But, trailers were loaded randomly and driven to the site where the trailers were off-loaded. Some unused pieces of steel can be seen in Fig. 6-2.

The superintendent had the opportunity to control the production, but did little to do so. The reduction in daily output is largely associated with steel deliveries. Steel was delivered to the site regularly, stockpiled at the site, and then

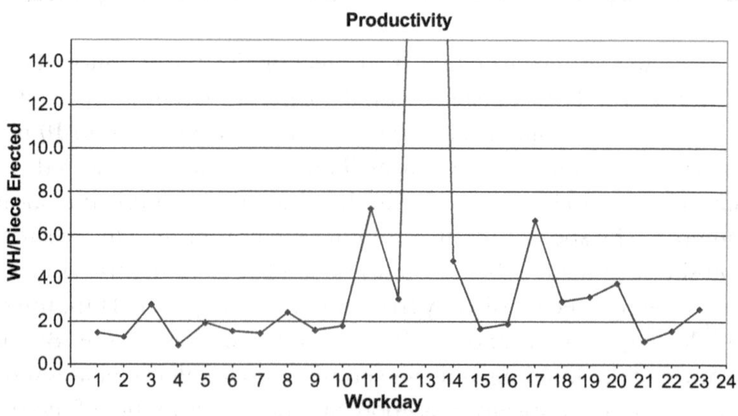

Fig. 6-34. Daily Productivity for Steel Erection, Food Science Building

erected from the stockpile. The problem with this approach is that off-loading and stockpiling took two to three members of the crew, leaving the other members of the crew with little or nothing productive to do. So daily output was greatly reduced on days when steel was delivered, which was done frequently. The superintendent failed to take advantage of the opportunity to erect steel directly from the delivery truck (Principle 4.1).

Based on 70 work hours/day and 1.20 work hours/piece, a daily goal of 58 pieces/day is calculated. The crew came close to this goal on only four workdays. The cumulative crew productivity was 2.06 work hours/piece compared with 1.32 work hours/piece on the Greenwich Court project described in Case Study 1. The cumulative steel erection productivity on the Food Science Building was about 60% worse than the Greenwich Court project.

References

Han, S. Y., and Thomas, H. R. (2002). "Quantification of labor inefficiencies—A case study." *Proc., Triennial Symp, Organization and Management of Construction*, CIB, Cincinnati.

Riley, D. R., and Sanvido, V. E. (1995). "Patterns of construction-space use in multistory buildings." *J. Constr. Eng. Manage.*, **121**(4), 464–473.

Riley, D. R., and Sanvido, V. E. (1997). "Space planning method for multistory building construction." *J. Constr. Eng. Manage.*, **123**(2), 171–180.

Stukhart, G., and Cook, E. L. (1990). "Bar-code standardization in industrial construction." *J. Constr. Eng. Manage.*, **116**(3), 416–431.

Thomas, H. R., Riley, D. R., and Sanvido, V. E. (1999). "Loss of labor productivity due to delivery methods and weather." *J. Constr. Eng. Manage.*, **125**(1), 39–46.

Thomas, H. R., and Sanvido, V. E. (2000). "The role of the fabricator in labor productivity." *J. Constr. Eng. Manage.*, **126**(5), 358–365.

Thomas, H. R., Sanvido, V. E., and Sanders, S. R. (1989). "Impact of material management on productivity—A case study." *J. Constr. Eng. Manage.*, **115**(3), 370–384.

CHAPTER 7

Fundamental Principles of Workforce Management

Construction site operations are subject to many disruptions related to workforce management practices, and these disruptions can result in significant economic losses to the contractor. Most of the research conducted on construction labor productivity has focused on the causes of subpar productivity (Thomas and Smith 1990; Horner and Talhouni 1993; Thomas et al. 2002) and the quantification of labor inefficiencies (Hanna et al. 1999a, b; Thomas and Napolitan 1995; Thomas and Raynar 1997; Thomas 1999; Thomas and Sanvido 2000). There is considerable knowledge about why things go wrong, but there is limited published information about effective management practices to apply to avoid cost overruns and time delays associated with labor inefficiencies, especially as it relates to workforce management. Several important principles are related to how one should effectively manage a construction workforce at the crew level (Thomas and Horman 2006).

7.1 Efficient Workforce Management Practices

The insights in this chapter apply to workforce management deficiencies in all phases of a project where labor is being used. Some of the common workforce management deficiencies observed on numerous projects are the following:

- Not organizing the crew into multiple work teams,
- Insufficient time lag (buffers) between activities,
- Having no expectation of the daily crew production, and
- Not doing housekeeping or incidental work concurrently with high-value work.

7.2 Fundamental Principles

Table 7-1 summarizes some fundamental principles of workforce management. The principles are organized according to the categories of general, daily work schedule,

Table 7-1. Fundamental Principles of Workforce Management

Number	Principle
	General
1.1	Where possible, use a 4-10 work schedule.
1.2	Staff the activity with labor resources that are consistent with the amount of work available to be performed. This includes taking adequate account of the variability in the project.
1.3	Have a good termination or layoff policy at the crew and project levels.
1.4	In instances of uncontrollable variability in workload, more labor than planned may need to be applied rapidly to complete work in the required time frame.
1.5	Superintendents and supervisors should not pick up tools and perform work normally done by workers.
1.6	The concept of a working supervisor is not a good idea.
1.7	Share with the crew expectations in terms of production, work hours, time, and completion dates.
1.8	Each crew must be allowed to work at an unimpeded pace to maximize labor efficiency.
	Daily Work Schedule
2.1	Daily schedules should be planned to prolong the periods of the most productive work activity.
	Work Assignments
3.1	Make the primary focus of the crew's work directed to high-value work. Never stop working on high-value work.
3.2	Efficient material handling and timely deliveries are important for good productivity, especially on labor-intensive operations.
3.3	Work on low-value subtasks concurrently with high-value work.
3.4	Perform incidental and cleanup work concurrently with high-value work.
	Crew Structure
4.1	A crew should be viewed as a collection of flexible size work teams.
	Disruptions
5.1	Minimizing or eliminating disruptions improves performance.
5.2	Keep the work face area free of trash and clutter.
	Multiskilling
6.1	Create multiskilled crews with individuals who have different skills, and who, when combined, form the multiskilled crew.
6.2	Where concurrent multiskilling tasks are applied, crew assignments need to carefully consider the size of the teams, the production output to be expected for each team, and how many hours the teams should be given to accomplish their work.
	Preassemblies and Modules
7.1	Preassemblies and modules are an effective way to reduce the field labor component of the installation activity.

Table 7-1. Fundamental Principles of Workforce Management *(Continued)*

Number	Principle
	Symbiotic Crew Relationships
8.1	Where possible, avoid symbiotic activities.
8.2	There should be adequate time buffers between each activity.

work assignments, crew structure, disruptions, multiskilling, preassemblies and modules, and symbiotic crew relationships.

General Principles

Though there is limited information about the labor efficiency of workdays of different lengths, unpublished research suggests that labor efficiency is not impaired so long as the workday does not exceed 10 hours. Contractors and craft workers like to work a four-day, 10-hour per workday (4-10) work schedule. This schedule gives the workers a three-day weekend while providing the contractor with a makeup day (Friday) without having to pay overtime wages in the event of adverse weather or some other calamity. Saturdays can be used for overtime work, if necessary. Therefore, contractors should seriously consider using a 4-10 work schedule (Principle 1.1). Some union agreements may preclude the use of a 4-10 schedule.

Another important advantage of the 4-10 schedule comes from recognizing the existence of daily startup and shutdown times. There are four startup and shutdown times that occur each day, at the beginning and end of the shift and just before and after lunch. These times are notoriously unproductive because the crew is getting organized, becoming acclimated to their daily work assignment, putting away tools, and cleaning up. With a 4-10 schedule, there are 16 unproductive periods each week. With a 5-8 schedule, there are 20 such periods. The advantage of the 4-10 schedule is obvious.

Contractors should staff an activity with labor resources consistent with the amount of work available to be performed (Principle 1.2). This principle seems trite, but it is often violated. How many times has the reader seen a six-person crew with two persons working and four persons watching? Such a situation is shown in Fig. 7-1. In the figure, six craftsmen have been assigned to do the work that can probably be done by four. Project engineers need to be sensitive to this situation as it occurs often on a construction site, and it costs the contractor money. Most contractors follow Principle 1.2 on a sitewide or craft basis, but less so on a crew basis. Many instances have been observed where full-sized crews have insufficient work to perform. If the other crew members do any work, they commonly perform "busy" or "filler" (incidental) work. However, if sufficient work is not available, the superintendent should reduce the crew size or divide the crew into teams. This should be done within the context of the labor agreement and a flexible labor management strategy,

Fig. 7-1. Overstaffed Work Assignment

as explained in the following. Also, care is needed not to adversely affect worker morale in these instances.

On one $45 million research facility and classroom project, the steel erection activity was monitored (see Chapter 6, Case Study 4). As the productivity suffered and the schedule slipped, the merit shop contractor increased the crew size from seven to nine to 10. This strategy simply meant more workers standing around and watching. The contractor's violation of Principle 1.2 cost the contractor a minimum of 800 work hours or $25,000–30,000 (at $35/h burdened) on what was a relatively straightforward project. The site was spacious and weather was not a problem, except for two days. The main problems were (1) not erecting steel directly from the delivery truck and (2) the addition of more ironworkers.

Too many workers can be a common occurrence at the end of an activity or project because superintendents are often reluctant to lay off workers (United Nations 1965). This work is made more difficult by excessive piecemeal, punch list, remedial, and incidental work. Fig. 7-2 shows the daily productivity of a structural steel erection activity. The eight-person ironworker crew was organized into four teams; one team was responsible for the initial erection of the steel members. Beginning on workday 18, the erection of steel pieces was finished, and the crew was overstaffed by about a third. Overstaffing is known to be the cause because on workday 22, the crew size was reduced from eight to two, and the productivity returned to about 1.1 wh (work hours)/piece. This can be clearly observed in Fig. 7-2. There were no other disruptions observed during this time frame. The contractor had no other project in the area where these workers could be sent. On this project, the inefficient work hours caused by overstaffing cost the subcontractor more than $6,000 (at $35 burdened).

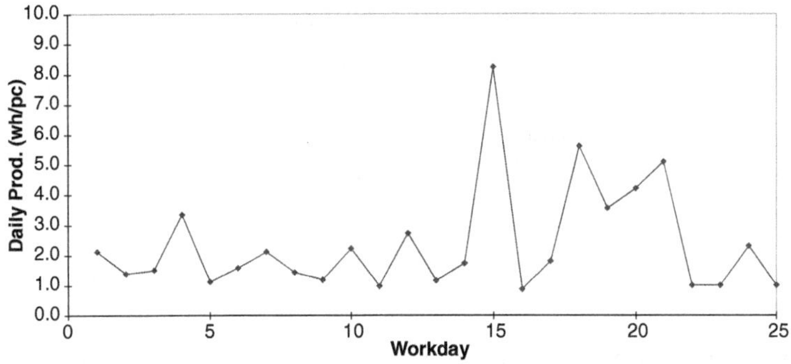

Fig. 7-2. Daily Productivity of Steel Erection Activity, Wartic Hall, Pennsylvania State University, State College, PA

A contractor needs to have a good termination or layoff policy at the crew and project levels (Principle 1.3). Being overstaffed at the end of a project is particularly problematic if a contractor is waiting for another project in the area to start or to progress to the point where the excess workers are needed. It has also been observed that some superintendents are reluctant to release workers because of the difficulty in forecasting future labor demands. Engineers need to develop robust methods of forecasting labor demands. These issues are particularly problematic during schedule acceleration (see Chapter 12) and can have serious cost implications.

In some instances, crews may need to be sized at a level above normal because work is fed to them erratically. This occurs when disruptions and events like late changes, delays by other contractors, and unforeseen site difficulties affect the project but completion dates are held rigid (Horman 2000). In instances of uncontrollable variability in workload, more labor than planned may need to be applied rapidly to complete the work within the required time frame (Principle 1.4). However, this strategy can be costly. The situation may be exacerbated by a tight labor market.

It is not uncommon for superintendents and supervisors to have worked their way up through the trades. It is hard for them to completely abandon their tools. But, superintendents and supervisors should not pick up tools and perform work normally done by craft workers (Principle 1.5). The superintendent's and supervisor's job is to plan and to anticipate problems, foresee shortages, schedule equipment, locate and assemble materials, and coordinate with other trades. These tasks cannot be done effectively if superintendents and supervisors are working with their tools. Thus, the concept of a working supervisor is not a good idea (Principle 1.6).

Sharing with the crew management's expectation of production goals is a good idea. If communicated, the goal is almost always met, as long the goal is realistic and management works with the crew to achieve the goal (Principle 1.7) and to remove obstacles. Occasionally, a good tactic is to ask the crew to provide the goal. It is almost always higher than management's expectation, and if the crew sets the goal, it is likely to be met. It follows that it is important to know what a crew can produce and to take appropriate action if production goals are not met.

Importantly, to maximize labor efficiency, each crew must be allowed to work at an unimpeded pace (Principle 1.8). Fig. 3-17 conveys a situation where this principle should be considered. Time lags are important to achieving production goals.

Daily Work Schedule

The intensity of the work (and output) varies throughout the workday. The most productive times occur around 10:00 a.m. and 2:00 p.m. It follows that supervisors and superintendents should try to schedule breaks to avoid interruptions to these productive times. Daily schedules should be planned to prolong the periods of the most productive work activity (Principle 2.1). In a study of a structural reinforcement activity on a major industrial site, it was determined that the work intensity was generally as shown in Fig. 7-3a. Fig. 7-3b shows a hypothetical work schedule and planned breaks for an eight-hour workday that prolongs the period of highest intensity. The break time in the morning is shown for 9:00 a.m., lunch is at 11:30, and there is an unscheduled break in the afternoon. This optimized daily work schedule provides longer periods of time when the work is most intense in both the mornings and afternoons.

Work Assignments

High Value Work—A construction activity usually consist of multiple subtasks. The relative labor effort applied to each subtask needed to produce a single unit of output is described by rules of credit. Table 7-2 shows example subtasks and rules of credit for concrete formwork and structural steel erection activities. Incidental subtasks are ignored. If, for example, on a steel erection activity, it is estimated that it will take 2,000 work hours, it should take approximately 200 work hours to tighten the structure (using the rule of credit of 0.10). As can be seen in Table 7-2, the erection activity in both instances requires the most labor effort. Subtasks with high labor content (not necessarily cost) are termed herein as "high-value" work. It is important to make the primary focus of the crew's work directed to high-value work at all times. Never stop working on high-value work (Principle 3.1). For example, if formwork erection is high-value work, then the work should be planned such that significant quantities of formwork are erected every day. While low-value work still needs to be done, it should be done concurrently with high-value work. Unfortunately, much of the time, low- and high-value work are done sequentially. Sequential practices drastically degrade crew productivity in most instances because low-value work often requires only a few workers, not the entire crew.

For instance, suppose a formwork crew of 12 carpenters working an eight-hour shift (96 work hours per day) has a goal of achieving a daily productivity of 0.08 wh/ft^2

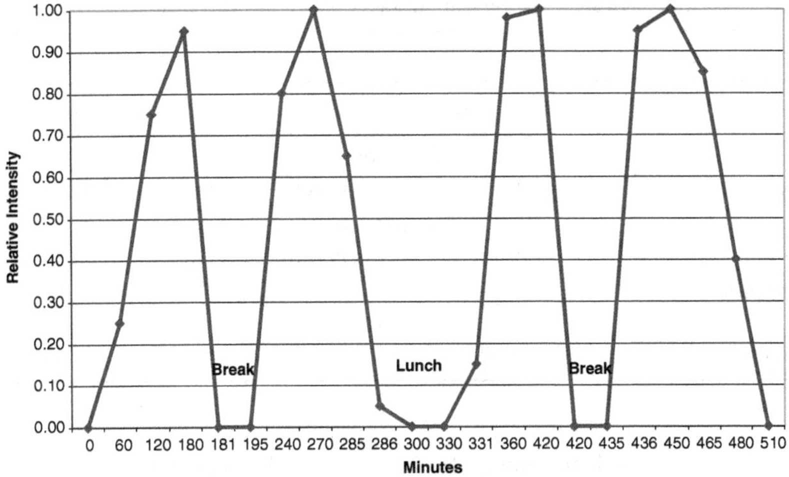

(a)

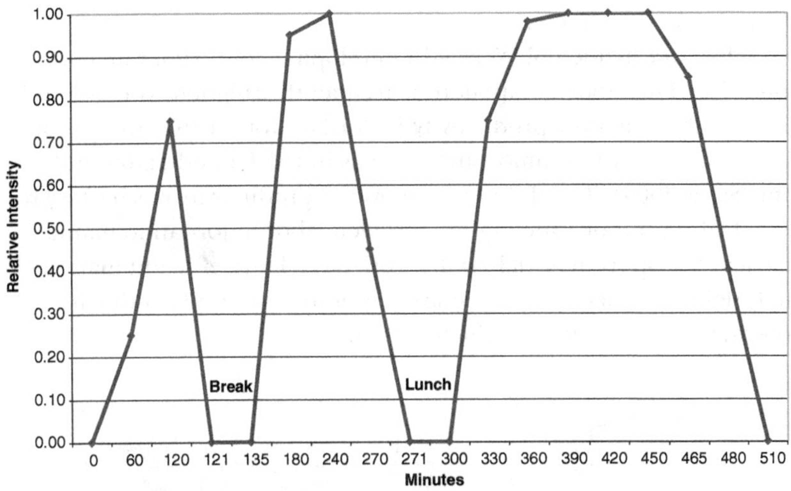

(b)

Fig. 7-3. Intensity for an Eight-Hour Workday. (a) Actual Intensity for an Eight-Hour Day, and (b) Intensity for an Optimized Work Routine

of wall formwork. To do so, the crew must be credited with 1,200 ft^2 of formwork each workday (96/0.08). Using the rules of credit in Table 7-2, the crew must erect 1,200/0.75 or 1,600 ft^2 to receive credit for 1,200 ft^2. Conversely, if stripping is the only daily activity performed (a low-value subtask), the crew must still be credited with 1,200 ft^2 to meet its budgeted performance. To get credit for 1,200 ft^2, the crew will have to strip an actual amount of 1,200/0.10 or 12,000 ft^2. There may not be 12,000 ft^2 of formwork on the site. This simple example shows the importance of Principle 3.1. Many construction activities, such as concrete formwork, steel erection, some reinforcement installations, and duct installation, have been observed where most of the daily work was spent on low-value work only. However, the good projects are the ones

Table 7-2. Typical Subtasks and Rules of Credit

Concrete Formwork		Structural Steel	
Subtasks	Rules of Credit	Subtasks	Rules of Credit
High-Value			
Erect	0.75	Erect	0.60
Low-Value			
Plumb and align	0.15	Plumb and align	0.30
Stripping	0.10	Tightening	0.10
Total			
Total	1.00	Total	1.00

where high-value work is the main effort every workday. Fig. 7-4 shows work where a duct crew has spent the entire workday erecting hangers (low-value work). No duct (high-value work) was erected.

Low-Value and Incidental Work—In developing countries, much of the work is labor intensive. The labor component is frequently assigned to material handling. Sweis, in a study of mason productivity in Jordan, found that the productivity for skilled labor (masons) was almost the same as in the United States and the United Kingdom (Sweis 2000). The difference in overall productivity was in the productivity of unskilled labor. It took much more unskilled labor in Jordan to manually unload, stockpile, and transport materials to the work face. Thus, efficient material handling and timely deliveries (see Chapter 6) are important for good productivity, especially on labor-intensive operations (Principle 3.2).

Fig. 7-4. Workers Have Done Hanger Erection Only this Day, Smeal College of Business, State College, Pennsylvania

In the erection of a precast parking garage, delivery of precast components was scheduled down to 15-min intervals (Fig. 6-32). Deliveries were timely, and labor productivity was good, except when the crane was moved. Fig. 6-33 shows consistent productivity of this operation (see Chapter 6, Case Study 3). The crane was relocated on workdays 11–13.

It is recognized that subtasks such as stripping of formwork and tightening of bolts on steel erection activities are necessary. Additionally, there are other incidental tasks that need to be performed, like cleaning of forms, site cleanup, and others. However, there is always time during the workday for these subtasks to be performed by a small team of workers. For instance, two or three carpenters could be taken from the work in Fig. 7-1 to perform these tasks.

Consider another example involving a crew of eight carpenters. For the erection of wall formwork, the following work requirements for each subtask is assumed: erection of gang forms: four carpenters; erection of modular forms: four carpenters; bracing, plumbing, and alignment: two to three carpenters; stripping: two to three carpenters; and bulkheads: two carpenters. If during the day, the work required includes erection of modular panels and installation of bulkheads (both subtasks combined may require six carpenters), then there are theoretically two carpenters available to perform low-value or incidental work. This situation may last for several hours. The underused members of the crew can perform stripping, cleaning of forms, or site cleanup while formwork erection (bulkheads and modular forms), which are high-value subtasks, are still being performed by the other crew members. Thus, an important principle is to work on low-value subtasks concurrently with high-value work (Principle 3.3). There is always some time during the day to do low-value work. If low-value work is not available, then alternate work assignments should be made, but the crew should be fully used.

Fig. 7-5 is a hypothetical daily work schedule for an eight-person crew of carpenters installing wall formwork. The schedule is divided into eight hourly work segments. The formwork subtasks are listed at the left of the figure along with the number of carpenters required to perform each subtask. As can be seen, during parts of the workday, a team of two carpenters is available for cleanup or other low-value or incidental work. What is important to avoid is the situation shown in Fig. 7-1 where six carpenters have been assigned the work that could have probably been done by four.

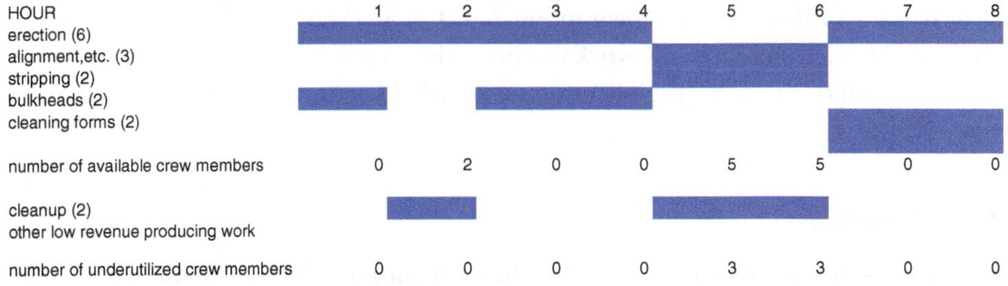

Fig. 7-5. Hypothetical Daily Work Schedule

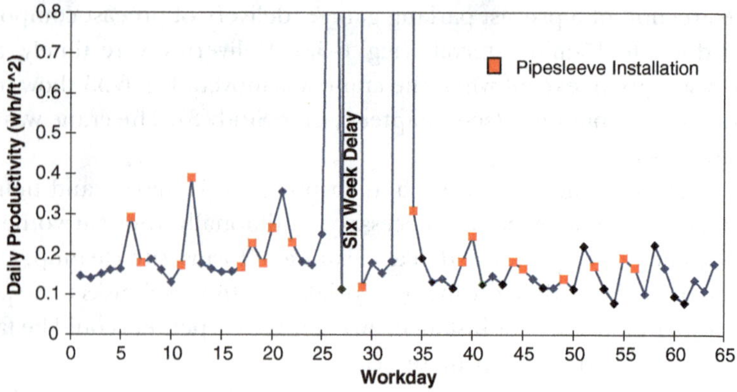

Fig. 7-6. Labor Productivity on a Wastewater Facility Showing the Effect of Low-Value Work

In yet another example, Fig. 7-6 shows the productivity of a wall formwork crew on a wastewater facility. The work included both wall formwork (gang formwork) and the installation of pipe sleeves through the walls. The crew size was six carpenters. Erection of gang forms required all six crew members, but the pipe sleeve subtask took two to three carpenters. The contractor chose to work four to five days on nothing but erecting wall formwork, then two to three days on nothing but installing pipe sleeves (low-value work), so the subtasks were done sequentially. Almost 30% of the workdays involved the installation of pipe sleeves of various sizes. On almost all of these days, the productivity is worse (by about 50%) than days when pipe sleeves were not installed. The supervisor assigned the entire crew to do this low-value work when maybe only a portion of the crew would have been justified. The supervisor violated Principles 3.1 and 3.3.

A related principle is to perform incidental and cleanup work concurrently with high-value work (Principle 3.4). Another example is worthwhile to illustrate this important principle. An efficient framing crew was observed during the partition wall construction of a unit of condominiums and townhouses. Unfortunately, the crew deferred the incidental task of cleanup until the production work was almost complete. At the end of the activity, it took three workdays to clean the site, thereby eroding much of the profit generated by the productive crew. The productivity of this crew is shown in Fig. 7-7. On workdays 16–18, the productivity degraded because most of the work was on incidental (cleanup) work. On workdays 1 and 2, most of the crew did low-value work only. "Clean as you go" is an important principle.

Crew Structure

Because it is always desirable for a crew to perform low-value work, incidental work, cleanup work, or be given alternate assignments concurrently with high-value work,

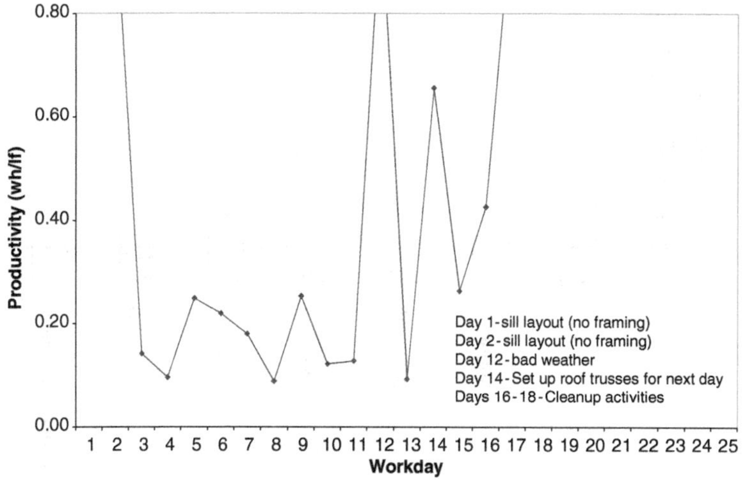

Fig. 7-7. Productivity of the Framing Crew

it is important that the crew be viewed as a collection of flexible size work teams. A few subtasks are likely to require all crew members, i.e., gang form erection, but many subtasks, like bulkheads, require a small subset of the crew. The teams need to be flexible in size because the work hour requirements vary with each subtask (Horman and Kenley 1998; Horman 2000). Therefore, a crew should be viewed as a collection of flexible size work teams (Principle 4.1).

A phenomenon occurs on many projects when the project is around 85–95% complete. The character of the work shifts away from bulk installation to that of completing systems, making systems operable, and punch list or finishing activities. The work before this changeover is bulk installation, and it requires a comparatively larger crew than the finish work. Project engineers need to be sensitive to this changeover work because from this point forward, the crew should be divided into small work teams, often consisting of only one or two workers. Not dividing the crew into teams will lead to a violation of Principle 1.2.

Disruptions

In a study of formwork on six bridge sites in central Pennsylvania (Thomas et al. 2002), the disruptions affecting the work were observed and recorded, and the inefficient work hours were calculated. The summary results in Fig. 7-8 show the total inefficient work hours by category of disruption. As can be seen, labor, rework, weather, and equipment were the leading causes of labor inefficiency. Thus, minimizing or eliminating disruptions reduces the variability in labor productivity and improves overall labor performance (Principle 5.1).

It is also important to keep the work face area free of trash and clutter (compare Fig. 6-4 to Figs. 6-9 and 6-20 to 6-22). Therefore, keep the work face area free of trash and clutter (Principle 5.2) at all times.

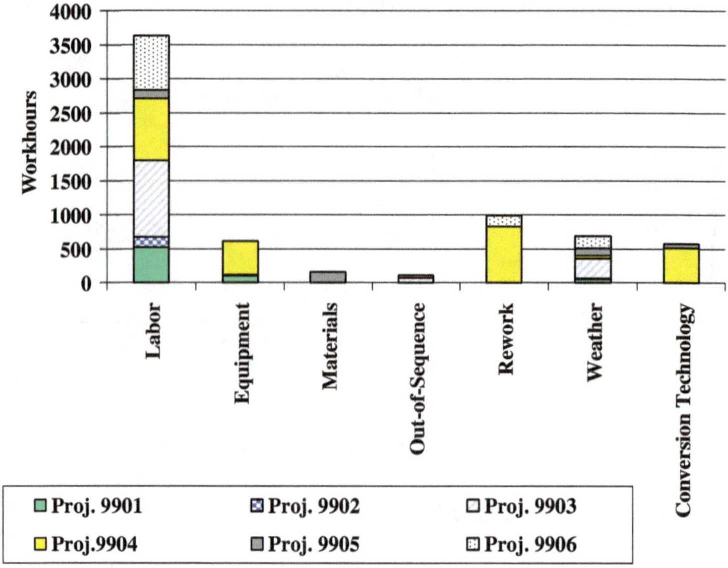

Fig. 7-8. Summary of Inefficient Work Hours by Resource

Multiskilling

Multiskilling refers to a crew's ability to perform multiple tasks that would ordinarily be assigned to several specialty crews (Burleson et al. 1998). Multiskilled crews can be created with differently skilled individuals that when combined form the multiskilled crew (Principle 6.1). An example of multiskilling is when one crew erects both concrete formwork and installs steel reinforcement. On workdays when both activities are in progress, the challenge is to allocate an appropriate portion of the crew to each task. To do so, one must assess the amount of work available for each task and base the assignments on what is considered to be an acceptable output for the workday. Alternate assignments can be made for parts of the day. Union labor agreements may preclude the use of multiskilled crews.

As an example of effective multiskilling, consider Fig. 7-9, which is the productivity curve of a crew erecting a concrete basement wall. The multiskilled crew performed formwork erection, steel reinforcement installation, and concrete placement. The superintendent moved crew members from one task to another based on the work available. He made sure that the teams were sized so as not to impede the progress of succeeding activities, and he also applied time lags. On workdays 4 and 11, alternate assignments were made (these work hours were included in the productivity calculation for that day). The results from this project indicate that the productivity for this crew was good. Use of multiskilled crews like this one requires constant monitoring to ensure that the teams are properly balanced. This requirement may prove to be burdensome.

In a less successful application of multiskilling, carpenter crews on four bridge projects erected abutments, foundations, piers, and pier caps. The crews performed

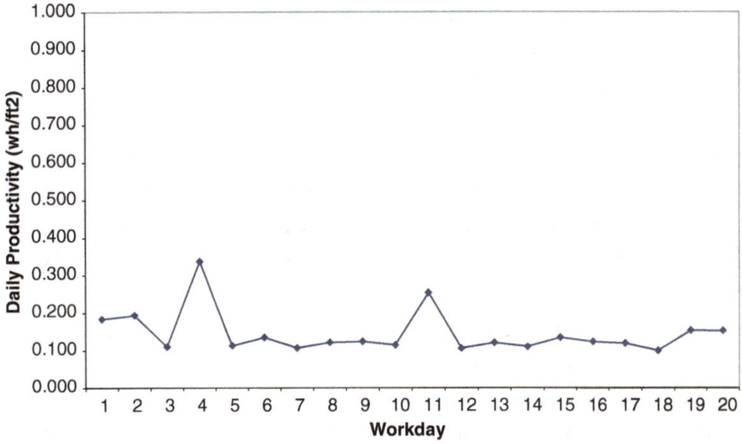

Fig. 7-9. Daily Concrete Formwork Productivity Showing Successful Use of Multiskilling, Alumni Center, Pennsylvania State University, State College, PA

formwork erection and placed concrete. Fig. 7-10 shows the crew productivity on the four projects (vertical bars) for the days in which concrete was placed. The figure also shows the size of the concrete placement (solid line). In general, the daily productivity improved as the size of the placement increased. For concrete placements where crew work assignments were split between concrete placement and formwork erection, the contractor seems to have had difficulty in properly allocating the appropriate number of individuals, especially for smaller placements (<100 cubic yards [cy]). Ensuring that sufficient work was available for both work teams and/or establishing expectations for the output and time frame for each team is not always an easy task, but it must be done to achieve consistent crew performance. From these data, it is concluded that where multiskilling is applied, the crew

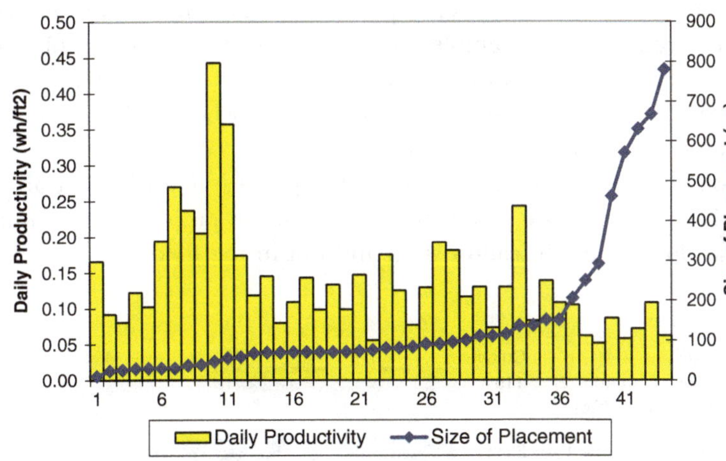

Fig. 7-10. Productivity versus Size of Concrete Placement

supervisor needs to carefully consider the size of the teams, the production output to be expected for each team, and how much time the teams should be given to accomplish their work (Principle 6.2).

Suppose that a crew consisting of six craft workers is to place 25 cy of concrete. (This is a multiskilled crew doing formwork and concrete placement.) The abutment formwork (modular) productivity budget is 0.090 wh/ft^2. If the concrete placement budget is 0.63 wh/cy of concrete, then the work hours needed to place the concrete and still be within budget is 15.75 or 16 work hours. If the placement is scheduled to take four hours, then four of the six crew members should be assigned to this work. The remaining two workers can be assigned to erecting modular formwork. In four hours, the smaller team of two should erect 119 ft^2 of abutment formwork $[(2 \times 4)/0.09/0.75]$. The rule of credit of 0.75 for erecting formwork (see Table 7-2) has been applied in this example. After four hours, the teams can then be merged to perform additional high-value work, say gang formwork. A hypothetical work schedule for the crew for the first five hours of the day is given in Fig. 7-11. If this goal is achieved, then the formwork and concrete budgets will be met.

A downside of a multiskilled crew is that teams may not be assigned to take advantage of the unique skills of individuals. If assignments are improperly done, productivity may suffer. In the previous example, the individuals most skilled in modular formwork erection should not be assigned to concrete placement.

Preassemblies and Modules

One way to minimize the amount of work crews do in awkward or unfavorable work face conditions is to preassemble components into subassemblies and erect the subassembly into position as a single unit. The assembly of column reinforcement cages is often observed. However, contractors can make wider use of this practice, and there are usually labor efficiency advantages in doing so. Han and Thomas observed that a crew installing duct required about 50% fewer work hours per foot when several sections were assembled on the floor and the subassembly was erected as a unit (Fig. 2-2), compared with the erection of sections one at a time (Han and Thomas 2002). Using preassemblies and modules may be an effective way to reduce the field labor component of the installation activity (Principle 7.1).

Fig. 7-12 shows the partial assembly of a support frame for a curtain wall. The preassembly yielded positive productivity gains on this project because much of the welding was done in a fabrication shop and not in the field.

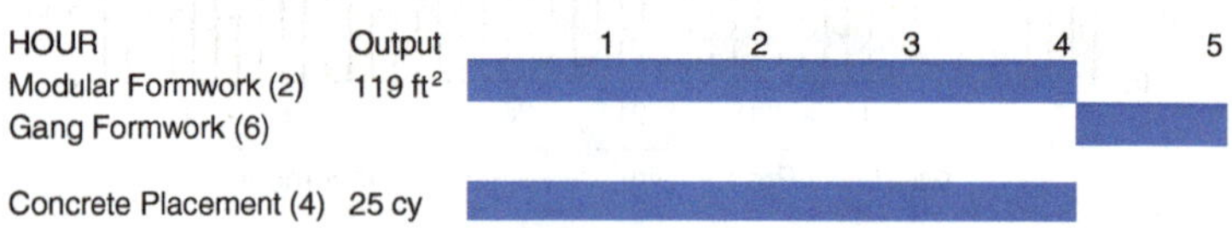

Fig. 7-11. Hypothetical Work Schedule

Fig. 7-12. Support Frame Subassembly for Curtain Wall, Dickinson Law School, Carlisle, Pennsylvania

Symbiotic Crew Relationships

Symbiotic work is cooperative work that is done by several crews on the same component or item of work. The work is passed from one crew or subcontractor to another and back again.

An example illustrates this concept. For the erection of a partition wall in an office building, a framing crew erects the wall frame, followed by an electrical crew that does rough-in electrical. Separate insulation and drywall crews complete the wall before the electrical crew returns to do the finish electrical work. The work of the electrical crew is symbiotic or cooperative, whereas the work of the framing crew is not. The framing crew can work at its own pace, whereas the pace of the electrical crew is likely to be dictated by others.

The problem with symbiotic work is that the individual, team, or crew doing symbiotic work is frequently in a standby mode, waiting for the work to be passed back to them. Their pace of work is dictated by others. Additionally, the individuals, teams, or crews doing the symbiotic work may be constantly changing work locations, which in itself is inefficient. Symbiotic work should be avoided where possible.

Project monitoring activities have shown that managing symbiotic activities can be difficult. Symbiotic crew relationships can sometimes be avoided by using prefabricated components. Also, schedule constraints can shift some activities from sequential to symbiotic (Thomas et al. 2004). Because of the management difficulties, it is important to avoid symbiotic relationships if possible (Principle 8.1).

Time Lags (Buffers)

A key productivity issue is whether a team or crew can establish its own pace of work. Where there are numerous activities and the production rates of activities are highly variable, a crew must always be allowed to work at its unimpeded pace. The only hope of establishing an efficient pace is to ensure that there are adequate buffers between activities (Principle 8.2).

Because of variable production rates and the exposure of construction activities to disruptions, it is important to maintain adequate stockpiles of materials and components. In a study of the productivity of steel reinforcement crews in Brazil, it was documented that the crew productivity was better with moderately sized stockpiles (Horman and Thomas 2005). Small stockpiles correlated with low performance, and large stockpiles had no performance advantages compared with moderately sized stockpiles.

Buffers are a way to lessen or eliminate the negative effect on labor productivity where there is high production variability in construction operations. Because it is unlikely that construction variability can be fully eliminated, construction managers must learn to manage the work under conditions of variability and learn to use buffers.

Types of Buffers—There are many types of buffers that allow teams and crews to work efficiently. Some common types are the following:

- Material stockpiles,
- Time lags,
- Variations in equipment and crew sizes,
- Subassemblies and modules,
- Alternate work assignments, and
- Others.

Probably the most common type of buffer is the time lag.

Schedule Acceleration—A significant problem with schedule acceleration is the management of buffers. Because there is such a rush to get the work done, many operations begin with little or no time lag between operations.

Fast and Slow Operations—Buffers allow for operations with varying rates of daily production to proceed efficiently. The need for time lags is illustrated in Fig. 3-17, which shows hypothetical velocity charts of an excavation and wall formwork operation on SCMB. Fig. 7-13 shows similar velocity charts for the completion of pipe racks on a refinery project. The pipe rack work shown in Fig. 7-13 involves the erection of steel pipe racks (frames) followed by the erection of large-bore pipe on the lower level of the steel frame, small-bore pipe on an intermediate level, and the installation of cable tray on the top level of the steel frame. For safety reasons, the work must proceed from the bottom level of the rack to the top level. Each of the four activities shown in Fig. 7-13 has a unique production rate. The slowest operation is large-bore pipe, and the fastest is the cable tray. Small-bore pipe is also a relatively fast operation. If small-bore pipe is started too soon after the start of large-bore pipe, then the crew doing the small-bore pipe will have to slow down or wait for the large-bore work to be done (lack of work in progress). Their work will be impeded unless there is sufficient time lag or time delay in starting their work. The same can be said about starting the cable tray too soon after the small-bore pipe. Each crew must be allowed to work at an unimpeded pace to maximize labor efficiency (Principle 8.1).

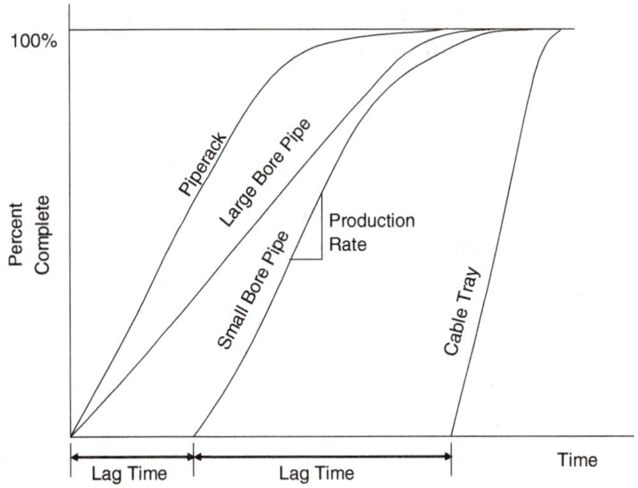

Fig. 7-13. Hypothetical Example Showing Differing (Unimpeded) Production Rates

High Production Variability in Preceding Operations—Time buffers are also necessary where the output of a preceding crew is highly variable (not just slow). Fig. 7-14 shows the daily quantities of structural steel erected on a commercial building project arranged in descending order. Note that more than 30% of the work was done in less than 15% of the elapsed time. This figure is typical of the output variability of many activities on all projects, regardless of the labor performance. Fig. 6-29 shows the productivity on the same project, and it was suggested that this was a good project. Changing from one component to another (columns to beams) also contributes to output variability as the production rates vary. For instance, the daily output of

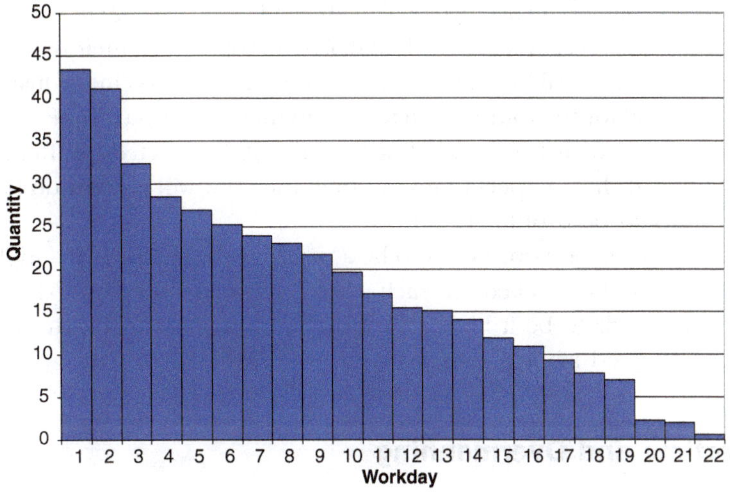

Fig. 7-14. Variable Output of Structural Steel, Greenwich Court, State College, Pennsylvania

concrete formwork measured in square feet varies significantly as the work proceeds from columns to beams to slabs. Disruptions also increase variability.

7.3 Expectations

Performance expectations are important to managers. The project engineer simply has to know what a crew can produce in a given period. When will the activity be complete? Will I be over or under budget at completion? In a recent study of union interior system contractors where production goals were not communicated, it was determined that the average number of sheets of drywall hung each day by an individual union carpenter varied from 10 to 30. Clearly, some contractors had low expectations of what an individual drywall worker can do. Sometimes an effective strategy is to let the crew set their production goals and completion dates. The following simple example shows how goals can be established.

Expectation Example

Suppose a crew of eight carpenters is responsible for framing and drywalling. There is one group of four identical rooms containing a total of 2,280 ft^2 (or 570 ft^2/room) of wall area. The drywall sheets are 4 ft × 10 ft × 1/2 in. Using an eight-hour day, the supervisor plans to assign a four-person team (one person to each room) to do this work. How many boards per hour should this team hang, and how long should it take to complete these four rooms? The budgeted unit rate for drywall is 0.0085 wh/ft^2.

The total work hours budgeted are $2{,}280 \times 0.0085 = 19.4$ work hours. If the four-person team works eight-hour days, this work should take 19.4/4 or about 5 hours (0.625 workdays).

Each room contains an average of $570/40 = 14.25$ boards (40 ft^2 each). If each of the four carpenters erects 14.25 boards in 0.625 workdays, then their daily production would average $14.250/0.625 = 22.8$ or 23 boards (40 ft^2)/day/individual.

Thus, the goal for the four-person team is to finish in 5 hours. The expectation is that on average, 23 boards (4 ft × 10 ft × 1/2 in.) will be erected per full workday by each carpenter. If these expectations are met, the crew will remain on budget, and the project should be profitable.

Of course, the supervisor needs to be a partner in this effort. The right amount of drywall should be stacked in each room, as shown in Fig. 7-15, and good housekeeping needs to be a high priority. The environment shown in this figure is conducive to good labor productivity.

7.4 Overtime and Overmanning

A commonly held belief is that to increase production, one simply needs to work scheduled overtime or add more workers. There are many dynamics that keep this

Fig. 7-15. Good Drywall Erection Practices Showing Uncluttered Work Face

strategy from being effective. While it is true that one may be able to work scheduled overtime productively for short periods, the circumstances have to be just right for overtime to accelerate the schedule efficiently. The same is true for adding more workers. Eight workers cannot normally double the output of four. Some of the dynamics that inhibit efficiency are congestion, fatigue, stacking of trades, accelerated use of materials, shared equipment, and increased production of trash and waste.

7.5 Case Study 1—SP45, SP62, and SP73: Time Lags

Three high-rise apartment building projects from Brazil are used to illustrate the need for buffers. Statistics for these projects are summarized in Table 7-3. The table shows the type of building, number of floors, and footprint area of each building. The structural components analyzed were columns, beams, slabs, and stairs. The activity studied was the on-site fabrication and installation of steel reinforcement. Cutting and bending of longitudinal and transverse pieces comprise the fabrication

Table 7-3. Project Descriptions

Project	Type of Building	No. of Floors	Footprint Area (m^2)	Rebar Needed for One Floor (kg)	Crew Size	Observation Days
SP45	Residential building	21	12,500	10,089	4	95
SP62	Residential building	13	5,000	5,955	3	35
SP73	Residential building	14	5,252	6,428	4	41

activity. Preassembly, material delivery, and final installation comprise the reinforcement installation activities. Time lags relate to time lapses between the fabrication of a certain percentage of reinforcement and the installation of that same percentage. For example, when was 35% of the reinforcement fabricated, and how much later was 35% of the reinforcement actually installed? This time lapse is the time lag, measured in workdays. The time lag is also related to the size of the fabricated stockpile.

The labor productivity for reinforcement installation only on project SP45 is shown in Fig. 7-16. The curve shows a high degree of variability. There were many days when no work was done.

The amount of reinforcement fabricated as a percentage and the time lag in days were calculated using the cumulative progress curve in Fig. 7-17 for project SP45 and similar curves for the three projects. Table 7-4 shows the average time lag in days for the three projects. The daily time lag varied more on SP45 compared with the other two projects.

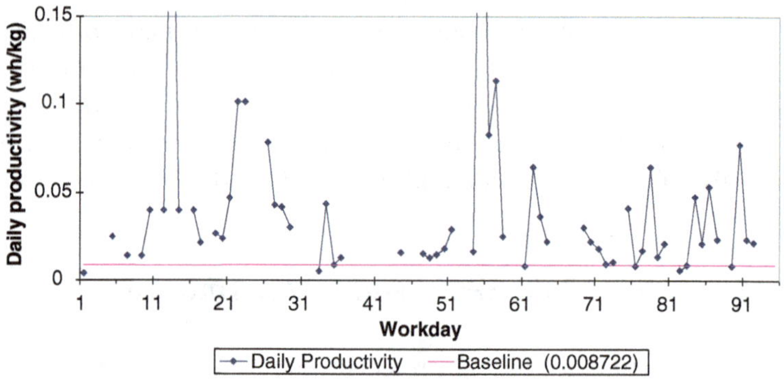

Fig. 7-16. Daily Productivity, SP45

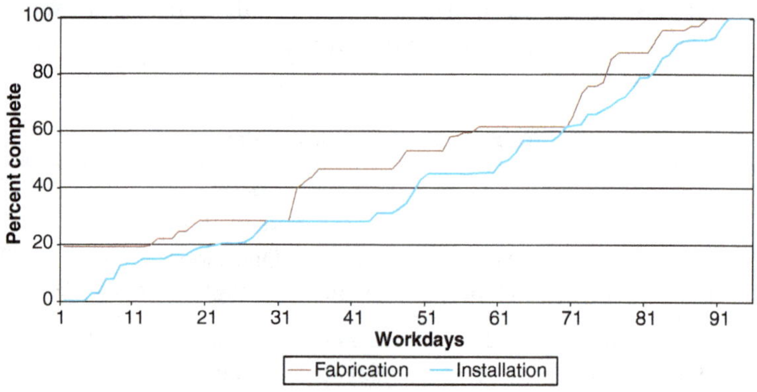

Fig. 7-17. Cumulative Progress Curves, SP45

Table 7-4. Average Time Lag

Project	Average Time Lag (Days)
SP45	8.6
SP73	6.4
SP62	5.2

Further analyses of all three projects were undertaken to see if the size of material stockpile was related to labor performance. On four days on SP45 when the stockpile was depleted and the time lag was zero, the productivity for the installation of steel reinforcement worsened or no work was performed. It seems that a zero stockpile buffer and smaller time lags are related to poor labor performance, while excess buffers did not improve performance.

Fig. 7-18 shows the conceptual relationship between labor performance and buffers for all three projects. With no buffer or too little buffer, labor performance degraded. An excess buffer has no redeeming virtue. Just-in-time material delivery (no stockpile) offers no productivity advantage. There is an optimal buffer that is unique for each operation. The goal of the manager is to identify the optimal size buffer.

The analysis of time lags and daily productivity on the three projects showed that where time lags were small, the productivity suffered adverse effects. For this operation, indications are that performance was best when the time lag was around four to five days. Each operation is unique, and four to five days is not a standard for all. However, the message is that managers, superintendents, and supervisors need to be sensitive to and make effective use of buffers and time lags so that crews can work at an unimpeded pace. If the buffer stockpile or surge pile is too small, workers may slow down their pace in anticipation of depleting their supply. This situation was observed

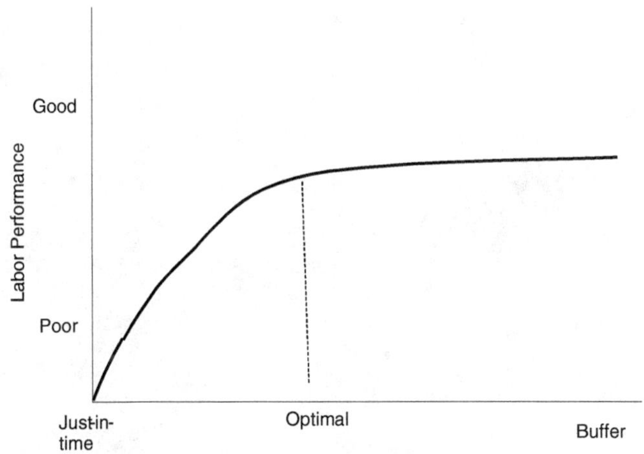

Fig. 7-18. Conceptual Relationship between Labor Performance and Buffers

on the Rider Building case study described in Chapter 6 (Fig. 6-28). Finding the appropriate buffer size is an iterative process. Managers should do the following:

1. Begin with a modest size buffer of one week (based on the budgeted productivity).
2. Observe the crew for any delays or negative effects (measure productivity using rules of credit).
3. Increase or decrease the buffer and repeat step 2 until there are no delays or negative effects.
4. Be sensitive to output variability and the ability of the crew to work at an unimpeded rate.

Some buffer is needed to permit superior labor performance. It is also clear from this study and other studies that variability must be managed in construction, although there is no clarity about the best way. It is clear that the concept of just-in-time material management needs to be weighed carefully. Other sources of variability in the fabrication and installation activity in this case study were not apparent from the data analyzed, and unfortunately, were unavailable.

7.6 Case Study 2—Bridges 28 and 29: Workforce Management Practices

These were a pair of identical cast-in-place reinforced concrete bridges on I99 (one northbound, the other southbound) in State College, PA. They were constructed by an experienced, nonunion, well-respected local contractor who has more than 35 years of experience in highway construction. The work monitored involved footers and abutment walls. The footers were rather massive, as shown in Fig. 7-19. The same

Fig. 7-19. Footers, Logan Branch Bridge, Bellefonte, PA

crew did the formwork, steel reinforcement, and concrete placement. When concrete was placed, the crew was sometimes augmented with additional laborers. The concrete placement crew size varied but averaged 10–13 craft workers.

The work was monitored for 57 workdays, during which 6,677 work hours were charged. The daily productivity is shown in Fig. 5-26. (Lower numbers denote better performance.) On about 20% of the workdays, the productivity was quite poor. Poor productivity was often observed to be followed by six to eight days of relatively stable productivity. The poor productivity days do not correspond to any particular work activity.

The daily diary was consulted in an effort to ascertain the causes of the productivity degradation, but no cause was found. The days of concrete placement were analyzed, and the results are summarized in Table 7-5. There were few disruptions to the concrete placement operations.

The data in Table 7-5 are revealing about the contractor's operation. In general, the footers were placed more efficiently than the abutments. There were no set days when concrete was poured. Larger pours were more efficiently done than smaller

Table 7-5. Pertinent Data on Workdays When Concrete Was Placed, Bridges 28 and 29

Workday	Component	Concrete (cy)	Work Hours	Productivity (wh/ft^2)	Comments
8	Footer	570	165[a]	0.058	Rain
9	Footer	80	70[a]	0.175	Subgrade elevation was incorrect
10	Footer	10	70[a]	0.166	Subgrade elevation was incorrect
11	Footer	250	77[a]	0.062	
12	Footer	83	32[a]	0.077	
19	Footer	780	246[a]	0.062	
22	Abutment	95	120[a]	0.181	
23	Abutment	30	114[a]	0.102	
28	Footer	136	240[a]	0.243	
33	Footer	90	190[a]	0.129	
34	Footer	630	290[a]	0.072	
36	Footer	150	104[a]	0.139	
39	Abutment	90	180[a]	0.192	
40	Abutment	150	130[a]	0.108	
43	Abutment	70	140[a]	0.109	
45	Abutment	137	80	0.083	
46	Abutment	80	70[a]	0.125	
49	Abutment	205	150[a]	0.105	
53	Abutment	115	130[a]	0.130	
54	Abutment	69	70[a]	0.145	Rain

[a]Other formwork was erected in addition to placing concrete.

ones, suggesting that the smaller pours may have been overstaffed, in violation of Principle 1.2. The contractor applied Principle 4.1 by concurrently placing concrete on smaller pours while erecting formwork with part of the crew.

However, a number of workforce management deficiencies were observed, and these are summarized in Table 7-6. The common theme related to workforce management issues was insufficient work to perform. On these days, the work consisted mainly of low-value and "incidental" work. Table 7-6 shows that this occurred on 15 workdays, or more than 25% of the time. On workdays 4, 5, 6, 15, 16, 18, 21, 41, and 48, the crew did mostly or exclusively low-value work. On workdays 17, 27, 28, 35, 39, and 52, incidental or "busy" work was done. Using an expected datum of 0.094 wh/ft^2, a total of 1,124 inefficient work hours were calculated because of workforce management issues or poor site planning, At \$35/h (burdened), this equates to almost \$40,000 wasted.

7.7 Case Study 3—Logan Branch Bridge: Workforce Management Practices

This bridge site also included north- and southbound spans, this time over Logan Branch on Interstate 99 in central Pennsylvania. A total of 75 workdays were monitored. A cursory review of the variability of the daily productivity in Fig. 7-20 suggests that this was not a particularly good project. The bridge was constructed by a contractor different from the one that constructed bridges 28 and 29. Nevertheless,

Table 7-6. Pertinent Data on Workdays When there Were Workforce Management Deficiencies, Bridges 28 and 29

Workday	Quantity (ft^2)	Work Hours	Productivity (wh/ft^2)	Comments
4	131.7	80	0.608	Insufficient work (Erect and brace)
5	68.8	80	1.163	Insufficient work (Brace only)
6	34.4	32	0.930	Insufficient work (Brace only)
15	133.2	70	0.526	Insufficient work (Brace only)
16	133.2	70	0.526	Insufficient work (Brace only)
17	0.0	80	—	Insufficient work (Site cleanup)
18	327.2	70	0.214	Insufficient work (Piecemeal work)
21	493.6	120	0.243	Insufficient work (Brace only)
27	0.0	22.5	—	Insufficient work (Setup for concrete pour)
28	987.8	240	0.243	Insufficient work (Wait to pour concrete)
35	1,228.6	250	0.203	Insufficient work (Site cleanup)
38	594.0	80	0.135	Insufficient work (Setup for concrete pour)
41	389.1	130	0.334	Insufficient work (Strip only)
48	120.0	130	1.083	Insufficient work (Brace and setup)
52	550.5	100	0.179	Insufficient work (Setup for concrete pour)

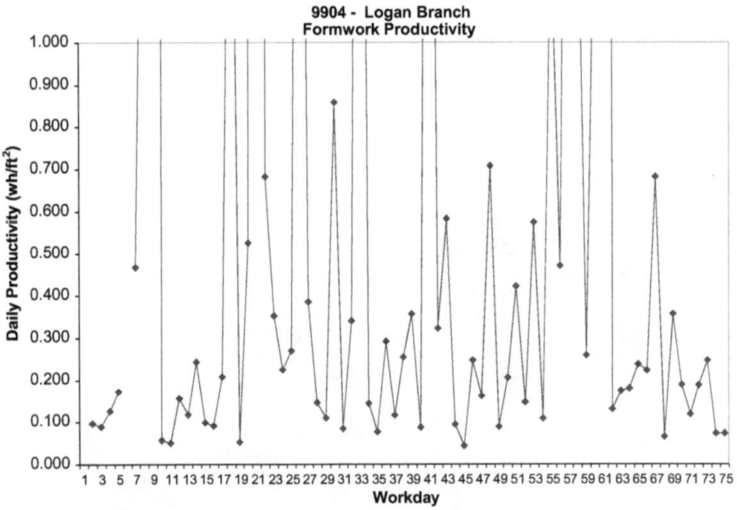

Fig. 7-20. Daily Productivity, Logan Branch Bridge

the contractor was experienced in highway construction and was a nonunion, well-respected local contractor with more than 45 years of experience in highway construction. The work monitored involved footers, piers, pier caps, and abutment walls. The piers were rather massive, as shown in Fig. 7-21. The same crew that did the formwork also did steel reinforcement and concrete placement. When concrete was placed, the crew was sometimes augmented with additional laborers. The crew size varied but averaged eight to 10 workers.

The work was monitored for 75 workdays, during which 9,798 work hours were charged. The cumulative productivity (in terms of equivalent wall formwork) was 0.174 wh/ft^2, compared with an expected productivity of 0.083 wh/ft^2. The daily productivity is shown in Fig. 7-20. (Lower numbers denote better performance.) Relatively consistent and average productivity is observed the first two weeks, then performance deteriorated and never stabilized thereafter.

Various disruptions occurred on this project, including rain, lack of equipment, rework, design errors, and material shortages. But of particular interest herein are workforce management deficiencies.

Workdays when concrete was placed were analyzed and are summarized in Table 7-7. In general, the placements were smaller than on Bridges 28 and 29, but the average crew size was also smaller. The smaller placements appear to have been overstaffed in many instances.

The daily diary was consulted, and it was learned that the crew productivity suffered because construction equipment was not available; it was shared with others. But the most significant event was a major design error. As the crew was preparing to place concrete, they observed that the footer for Pier 3 was mislocated by about 10 ft. This observation occurred on workday 20. Over the next seven to eight days, the crew had to disassemble the pier and footer steel, which were mostly #11–#18 bars,

Fig. 7-21. Piers, Logan Branch Bridge, Central Pennsylvania

Table 7-7. Pertinent Data on Days When Concrete Was Placed, Logan Branch Bridge

Workday	Component	Concrete (cy)	Work Hours	Productivity (wh/ft^2)	Comments
10	Pier	75	42	0.056	
15	Pier	70	70	0.099	
19	Pier	291	150	0.052	
37	Footer	100	60[a]	0.116	
39	Footer	56	100	0.357	Rework on pier 3
40	Footer	461	200	0.087	
50	Footer	39	40	0.205	Rain
54	Footer	667	360	0.108	
62	Pier cap	110	100	0.130	
65	Abutment	37	90	0.118	
71	Pier	68	80	0.118	
74	Pier cap	110	56	0.073	

[a]Other formwork was erected in addition to placing concrete.

disassemble the footing formwork, and reassemble all the components in another location. The crew received no credit for disassembly or reassembly. Therefore, its productivity suffered greatly until about workday 28. Unfortunately, the crew never recovered, and its productivity continued to degrade thereafter.

A number of workforce management deficiencies were observed. These are summarized in Table 7-8. The common theme related to workforce management was insufficient high-value work to perform. On these days, the work consisted mainly of incidental work in preparation for a concrete pour. Table 7-8 shows that this occurred on six workdays. On workdays 48, 56, and 67, the crew did some low-value work. On several workdays, they had nothing to do. Using an expected datum of 0.083 wh/ft^2, a total of 915 inefficient work hours were calculated because of workforce management issues or poor site planning, at \$35/h (burdened), this equates to about \$32,000.

7.8 Case Study 4—The Palmerton: Workforce Management Practices

The Palmerton project is a seven-story apartment complex with commercial space on the ground floor and two levels of underground parking. It was constructed in State College, Pennsylvania, in 2007–2008 at a cost of about \$15 million. The gross floor area is 133,000 ft^2, and the building height is approximately 82 ft. The unit cost was \$113/ft^2.

The foundation through the second floor is cast-in-place concrete. The remaining floors are masonry. All levels rely on precast floor planks. The building exterior is brick and synthetic stucco (exterior insulation and finish system, or EIFS). The completed project is shown in Fig. 7-22.

Table 7-8. Pertinent Data on Workdays When Workforce Management Deficiencies Were Observed, Logan Branch Bridge

Workday	Quantity (ft^2)	Work Hours	Productivity (wh/ft^2)	Comments
7	107.1	50	0.467	Insufficient work (setup for concrete pour)
8	21.4	42	1.961	Insufficient work (setup for concrete pour and cleanup)
9	0.0	60	—	Insufficient work (no work performed)
14	452.9	110	0.243	Insufficient work (setup for concrete pour)
18	110.2	220	1.997	Insufficient work
36	343.4	100	0.291	Insufficient work (setup for concrete pour)
41	38.3	120	3.130	Insufficient work (setup for concrete pour)
48	128.7	91	0.707	Insufficient work (brace and strip only)
53	174.4	100	0.573	Insufficient work (setup for concrete pour)
56	127.7	60	0.470	Insufficient work (brace and strip only)
67	64.6	44	0.681	Insufficient work (strip only)

Fig. 7-22. Artist's Rendition of the Palmerton Apartment Complex

The activity observed in this case study was the construction of underground reinforced concrete wall formwork. The walls were constructed using gang forms by a crew of 10 carpenters working five days per week. The gang forms were custom built from lumber and plywood in lieu of commercial formwork. The gang panels were constructed initially on the ground and thereafter erected and stripped as gang forms. This operation is shown in Fig. 7-23. Initially, the tower crane planned for this activity was not erected. The crew used a small (about 30-ton) crane. The crane can be seen in Fig. 7-24. Notice that the earthen ramp is preventing the completion of the reinforced concrete wall. Also shown in Fig. 7-25 is the partially erected tower crane. When the tower crane was finally erected, a power line passing through the middle of the site impeded its use.

From the outset, the formwork activity progressed slowly. No expectations of daily production were conveyed to the crew. Often crew members could be observed

Fig. 7-23. Assembly of Gang Forms

Fig. 7-24. Erection of Gang Forms Using 30-Ton Crane

Fig. 7-25. Tower Crane Being Erected Beneath a Power Line

as idle. At first, it could be argued that the crew was waiting for the tower crane to be erected. But even after it was erected, the slow pace of work continued. The wall formwork productivity on this project was awful. The cumulative productivity was calculated as 0.40 wh/ft^2, and the best daily productivity the contractor could consistently achieve was about 0.20 wh/ft^2. Previous research has shown that for a project of this simplicity versus complexity, a contractor should be able to consistently achieve a productivity of about 0.06–0.12 wh/ft^2.

The daily productivity graph for the first month is shown in Fig. 7-26. It shows no semblance of consistent performance until after the tower crane became serviceable. But even then, performance never approached 0.06 wh/ft^2. The activity was plagued by workforce management and congestion issues. The various subtasks were done

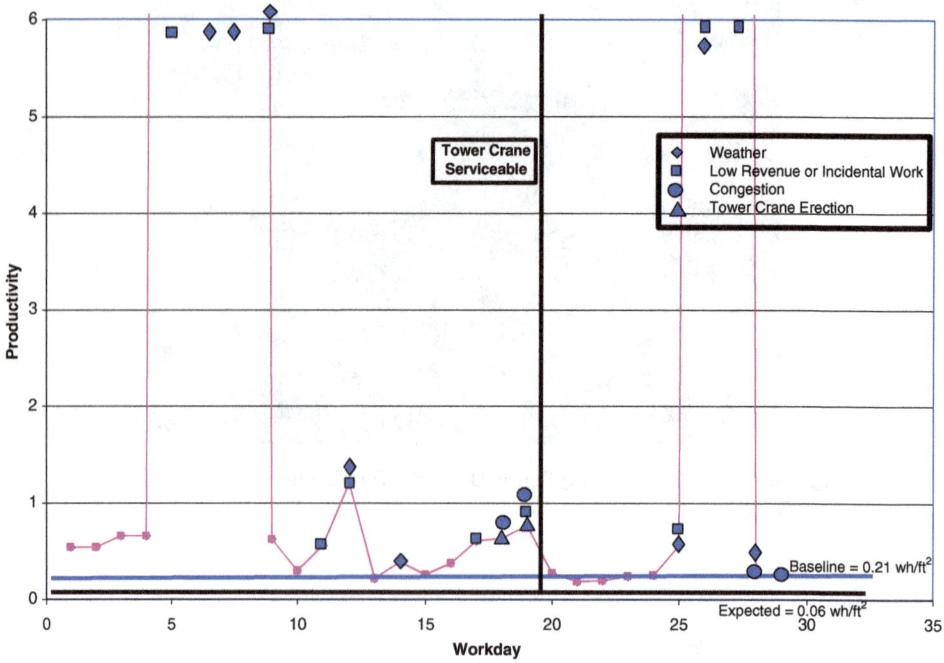

Fig. 7-26. Daily Productivity of Gang Formwork Erection

largely sequentially. On many workdays, the crew erected no formwork at all; most notably, little or no formwork was erected on workdays 5–8, 11–12, 17, 19, and 25–27. (On 11 of the 29 workdays, despite the fact that this was a critical path activity, little or no formwork was erected.) Instead, the crew did incidental or other work or did only low-value work. Disruptions related to workforce management became more common the longer the tower crane remained unserviceable.

An inefficiency analysis was conducted, and 474 work hours were chargeable to workforce management practices. At a burdened rate of \$35/h, this amounts to almost \$16,700 in just 29 days of work, or an average of \$572/day. There were an additional 956 inefficient work hours that could not be explained. It is likely that many of these inefficient work hours are the result of poor workforce management practices too. It should be obvious from this case study that poor workforce management practices can quickly erode profits. The overall labor overrun on this activity was calculated as 541%. Almost 1,800 work hours were charged during the observational time frame, whereas slightly fewer than 400 should have been needed.

7.9 Case Study 5—Smeal College of Business: Sequential Scheduling Practices

This case study is of the Smeal College of Business Administration Building on the Penn State University campus. This four-story, 210,000 ft^2 building was built in the

2004–2005 time frame at a cost of about $68 million. The building footprint was 46,000 ft^2, and the site area was approximately 324,000 ft^2. The construction schedule was unexpectedly shortened by three months to 21 months because the workers performing demolition of an existing structure surprisingly found asbestos. So the site was spacious, but the schedule was aggressive. The case study illustrates the effect of sequential scheduling practices and the consequences of not doing high-value work concurrently with low-value work. The completed project is shown in Fig. 7-27.

A construction manager agency mode of procurement was used. The general contractor was a local nonunion contractor. The activity monitored was partition wall framing. No sequence plan is known to have been developed.

Fundamental Principles

During the observation period, the following fundamental principles were applied: 1.2 and 4.1. These principles were not applied: 1.1, 3.1, 3.3, and 3.4. And these principles are not applicable to this project or their usage is unknown: 1.3, 1.4, 1.5, 1.6, 1.7, 1.8, 2.1, 3.2, 5.1, 5.2, 6.1, 6.2, 7.1, and 8.1.

Activity Description

The work reported in this case study is the framing of the interior walls and a small amount of exterior wall. The work involves layout followed by the installation of the upper track of the walls. Next was the installation of the bottom track and metal studs. The planned sequence of the work was to lay out and install the top track first. This was to be followed by fireproofing and then duct installation. The bottom track and studs were planned to be installed much later. Because of time limitations, only 30 workdays were observed, during which time only top track was installed. Relative to the top track, layout was low-value work, and track installation was high-value work. The top track that was installed during the observation period represented about 12% of the overall amount required.

Fig. 7-27. Smeal College of Business Administration

Contractor Operations

The work was largely done sequentially. The crew spent time laying out the top track. This step would be followed by installation of the top track. The sequential nature of the operation is shown in Fig. 7-28. The problem with this strategy is that when only layout was performed (a two-person operation), the remainder of the crew had little or nothing to do. When only track installation was performed, the crew may have been overstaffed, but this aspect is not analyzed. The crew size range was three to seven carpenters. There was likely much idle time. Only about 33% of the time was laying out and installing track, both done during the same workday.

The overall wall framing operation was symbiotic. The top track was installed first by a crew of carpenters. The partially completed wall was then passed to the fireproofing and duct crews before being passed back to the same crew of carpenters for installation of the bottom track and studs. Fig. 7-29 shows hypothetical velocity charts of these operations. The secret to planning these operations is twofold. First, there must be sufficient time lags (buffers) to keep fireproofing isolated from the other operations. Overtime or nighttime work may be a suitable alternative. Second, the framing crew must be sized so that the top track is completed when the bottom track is ready to begin.

Effects on Performance

The work was plagued by multiple factors. At times, the weather was unfavorable, and the building had been neither enclosed nor sealed adequately. Heating could not be effectively provided, and the trailer to provide warmth for the crafts was remotely

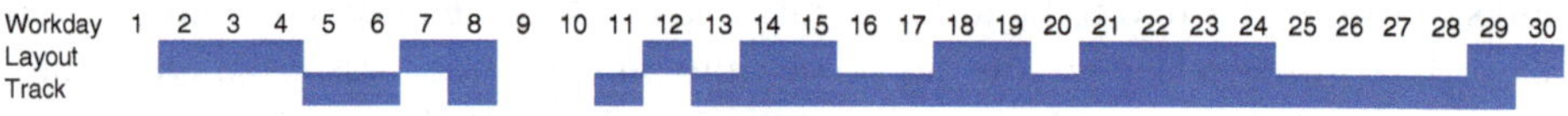

Fig. 7-28. Sequential Scheduling Practice

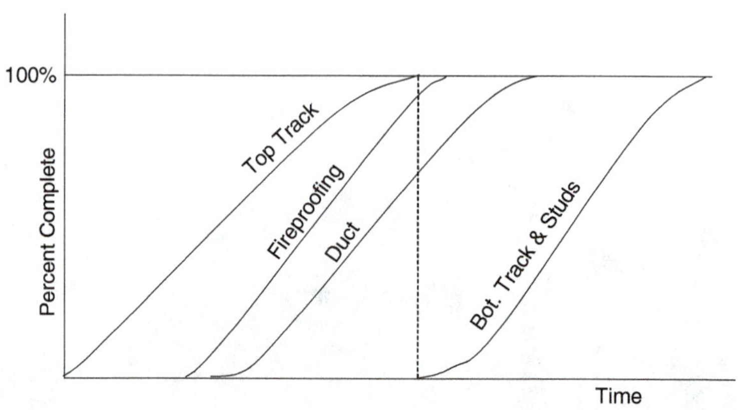

Fig. 7-29. Velocity Charts of the Framing Operation Showing Its Symbiotic Nature

located. The wind was also a negative factor. In all, there were five days that were cold and windy. There were three other days when it rained. The building had not been sealed, so water made its way into the building. In areas where there was standing water, cold overnight temperatures sometimes caused the water to freeze, precluding work in those areas the next day. The affected days are shown in Fig. 7-30. On days when layout was all that was done (a sequential scheduling practice), labor performance was especially poor.

The project was also plagued by sequencing problems. Initially, the work area was partitioned into six subareas. Unfortunately, there was never any semblance of a plan on how to progress from one area to another. The crew would work in whatever area was available. During the 30-day observation period, no work was done in areas five and six because the slabs in those areas had not been completed. Also, there was no obvious plan for progressing from one floor to another. The crew would work on whatever floor provided work. Sometimes the crew worked on multiple floors. The superintendent struggled to staff the project with labor resources consistent with the amount of work that was available (Principle 1.2). The highly variable daily crew size shown in Fig. 7-31 is indicative of the varying work available for the crew. The workdays when there were sequencing problems are not shown in Fig. 7-30 because the sequencing problems occurred every day.

On two days, the crew performance was adversely affected by a design error and late approval of shop drawings. On six other days, performance was affected by

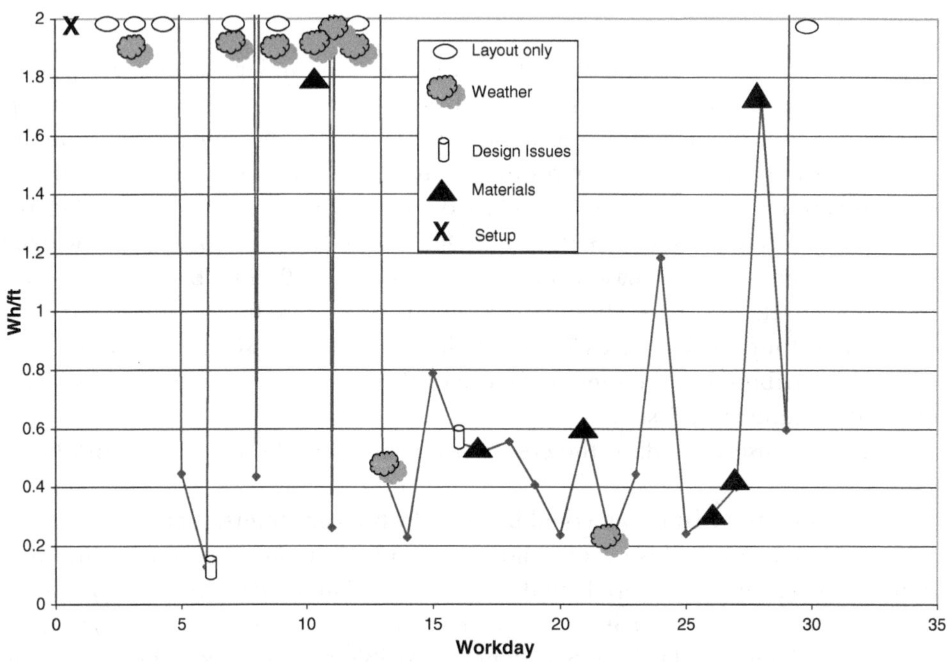

Fig. 7-30. Framing Crew Productivity, Smeal College of Business

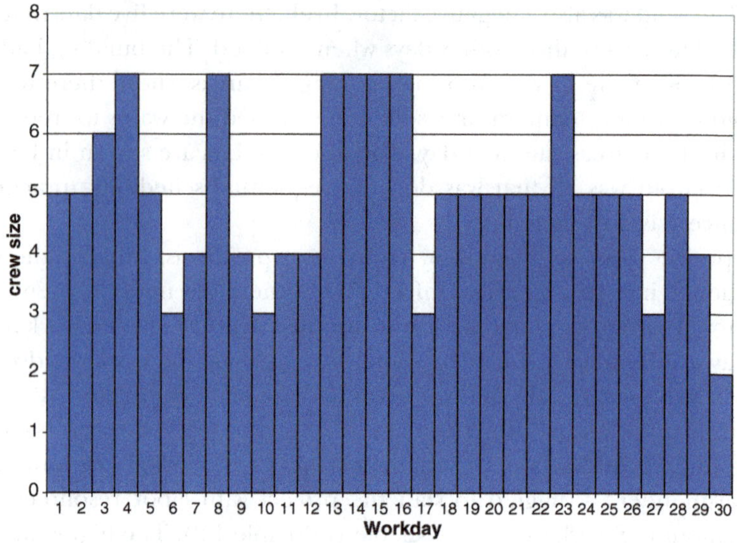

Fig. 7-31. Framing Crew Size, Smeal College of Business

material-related problems. The crew had preloaded materials, but the superintendent had apparently not ordered enough track because material deliveries were made three times.

Analysis

The variability of the productivity values in Fig. 7-30 indicates that the performance on this project was not particularly good. The best performance that the contractor could achieve was computed as 0.211 wh/ft. This value is based on five days of work in which the crew produced more than 40% of the 30-day output. The remaining 60% of output took 25 days. As can be seen in Fig. 7-30, the contractor rarely achieved better than 0.250 wh/ft. The work hour overrun was computed as 146% (based on a sustained rate of 0.211 wh/ft). At the rate of 0.211 wh/ft, the work should have taken 488 work hours. Instead it took 1,201 work hours, which yields 713 inefficient work hours. Most of this inefficiency can be attributed to the ineffective use of the workforce, which was exacerbated by the sequencing issue.

On two consecutive days, the crew did no work. On those days, 56 work hours were charged.

If one assumes that layout could be done by two carpenters, then on days when layout only was done, this would amount to 16 work hours. The remaining work hours that day are considered inefficiently used. There were six workdays when layout was the only work that was done. The total inefficient work hours for these six days was determined to be 128 work hours. At $35/h (burdened), this amounts to a loss of almost $4,500 because of not doing high-value work concurrently with the low-value work. If the 713 inefficient work hours are extrapolated to the entire job, the inefficiency loss is almost $208,000.

References

Burleson, R. C., Haas, C. T., Tucker, R. L., and Stanley, A. (1998). "Multiskilled labor utilization strategies in construction." *J. Constr. Eng. Manage.*, **124**(6), 480–489.

Han, S. Y., and Thomas, H. R. (2002). "Quantification of labor inefficiency." *Proc., 10th Triennial Symp. of Int. Council on Innovation in Building Construction,* Cincinnati, OH.

Hanna, A. S., Russell, J. S., Gotzion, T. W., and Nordheim, E. V. (1999a). "Impact of change orders on labor efficiency for mechanical construction." *J. Constr. Eng. Manage.*, **125**(3), 176–184.

Hanna, A. S., Russell, J. S., Nordheim, E. V., and Bruggink, M. J. (1999b). "Impact of change orders on labor efficiency for electrical construction." *J. Constr. Eng. Manage.*, **125**(4), 224–232.

Horman, M. J. (2000). "Process dynamics: Buffer management in construction." Ph.D. dissertation, Univ. of Melbourne, Parkville, Australia.

Horman, M. J., and Kenley, R. (1998). "Process dynamics: Identifying a strategy for the deployment of buffers in building projects." *Int. J. Logist. Res. Appl.*, **1**(3), **1**(3), 221–237.

Horman, M. J., and Thomas, H. R. (2005). "Role of inventory buffers in construction labor performance." *J. Constr. Eng. Manage.*, **131**, 834–843.

Horner, R. M. W., and Talhouni, B. T. (1993). *Effects of accelerated working, delays and disruptions on labour productivity,* Chartered Institute of Building, London, 40.

Sweis, G. J. (2000). "Impact of conversion technology on productivity in masonry construction." Ph.D. dissertation, Northwestern Univ., Evanston, IL, 187.

Thomas, H. R. (1999). "Schedule acceleration, workflow, and labor productivity." *J. Constr. Eng. Manage.*, **126**(4), 261–267.

Thomas, H. R., and Horman, M. J. (2006). "Fundamental principles of workforce management." *J. Constr. Eng. Manage.*, **123**(2), 97–104.

Thomas, H. R., Horman, M. J., and de Souza, U. E. L. (2004). "Symbiotic crew relationships and labor flow." *J. Constr. Eng. Manage.*, **130**(6), 908–917.

Thomas, H. R., Horman, M. J., de Souza, U. E. L., and Završki, I. (2002). "Benchmarking of labor-intensive construction activities: Lean construction and fundamental principles of workforce management." Rep. 276, International Council for Research and Innovation in Building and Construction, Rotterdam, Netherlands, 156.

Thomas, H. R., and Napolitan, C. L. (1995). "Quantitative effects of construction changes on labor productivity." *J. Constr. Eng. Manage.*, **121**(3), 290–296.

Thomas, H. R., and Raynar, K. A. (1997). "Scheduled overtime and labor productivity: Quantitative analysis." *J. Constr. Eng. Manage.*, **123**(2), 181–188.

Thomas, H. R., and Sanvido, V. E. (2000). "Quantification of losses caused by labor inefficiencies: Where is the elusive measured mile?" *J. Constr. Law Bus.*, **1**(3), 1–14.

Thomas, H. R., and Smith, G. R. (1990). "Loss of construction labor productivity due to inefficiencies and disruptions: The weight of expert opinion." Report to the National Science Foundation, PTI Rep. No. 9019, Pennsylvania State Univ., Pennsylvania Transportation Institute, University Park, PA, 181.

United Nations Committee on Housing, Building, and Planning. (1965). *Effect of repetition on building operations and processes on site,* United Nations, New York.

CHAPTER 8

Fundamental Principles of Activity Sequencing

An orderly sequence of construction operations is essential to a profitable construction project. Efficient labor performance demands a sequence of operations that promotes efficient methods and good workforce management practices. A thoughtful sequence can also lead to significant shortening of the project schedule (see Chapter 3).

There are two dimensions to sequencing: (1) The most commonly thought-of dimension is trade sequencing (see Chapter 9), that is, in a building project, how to plan for the efficient execution of multiple trades and subcontractors. Trade sequencing is frequently associated with stacking of trades and congestion. (2) A second dimension that is less widely discussed is the more general dimension of activity sequencing. While not totally isolated from trade sequencing, good activity sequencing can remove activities from the critical path (in a critical path method [CPM] schedule), shorten project schedules, and promote more labor-efficient operations on all projects, not just buildings.

This chapter presents some fundamental principles for sequencing of construction activities. By following these principles, contractors can be more cost-effective in executing their work.

8.1 Fundamental Principles

The analysis of the literature (Chapter 2) highlights the need for contractors to sequence their jobs to promote labor efficiency and for owners to have better guidelines to use in evaluating contractor schedules. Table 8-1 lists some fundamental principles for activity sequencing of construction operations. The principles have been intentionally left general so that they are applicable to a wide range of projects and conditions.

Submittals

Certain specialty items require long lead times for fabrication and delivery. In addition, there is the time required for submittal approval. It may take several weeks

Table 8-1. Fundamental Principles of Activity Sequencing

No.	Activity Sequencing Principle
	Submittals
1.1	Make schedule allowances for approval of submittals, fabrication, and delivery, especially for long-lead items.
1.2	Develop a submittal schedule showing due dates and enforce the schedule.
	Economies of Scale
1.3	Consider economies of scale for pertinent components and propose pertinent changes to the designer.
1.4	When applying Principle 1.3, avoid sizing components that require the use of a crane; size components to be handled manually.
	Eliminating Tasks
1.5	Sequence activities to eliminate work tasks.
	Avoiding Congestion
1.6	Sequence activities to avoid congested work areas.
	Managing Critical Path Activities
1.7	Sequence activities to minimize the number of critical path activities.
1.8	Complete critical path activities before starting noncritical activities.
	Shortening the Project Schedule
1.9	Where possible, do activities concurrently.
1.10	On slow activities, use multiple workstations.
	Utilities and Site Works
1.11	Do work early on site (underground and aboveground) utilities.
1.12	Backfill around the facility as quickly as possible.
1.13	Install permanent drainage facilities first.
	Enclosing and Sealing
1.14	Use permanent windows in lieu of a temporary (polyethylene) enclosure.
	Buffers (Time Lags)
1.15	Make use of buffers (time lags) between activities to allow crews to work at an uninhibited pace.

or more to get some submittals approved. After approval, items must be fabricated and delivered. If the item is needed early in the construction process, then timely submission of the submittal is imperative.

If the notice to proceed (NTP) is given shortly after the intent to award a contract is issued, the project completion date may not be advanced. Contractors should be using this time period to initiate the shop drawing process for items that are needed early, such as foundation steel or items related to site drainage. Shortening this time period does not necessarily shorten the fabrication process or accelerate project completion.

Occasionally, the components as detailed in the plans do not fit. Designers may instruct the contractor to "make it work on the shop drawings," or designers may say "this is means and methods." These situations may require the contractor to engage in design work, thus requiring more time in the fabrication process. A registered engineer may also be required. Such practices are viewed unfavorably by the authors.

If the start of the project is delayed by, say, a month, the contractor may lose his or her scheduled slot in the fabrication shop. The contractor may not be able to schedule another slot until two or three months later. Thus, a one-month delay in starting the project may result in a delay of several months in receiving fabricated items.

When rehabilitating or remodeling a facility, test walls are often required to match mortar, brick, and other features of the existing structure. Matching can be especially problematic, and multiple rejections are not unusual.

The submittal process is an important one and schedule allowances need to be made for selected items (Principle 1.1). This factor necessitates a submittal schedule with due dates (Principle 1.2). Unfortunately, Principle 1.2 is not followed on many projects. Two examples for buildings where long lead times may be needed are elevators and windows. Late arrival of either can have an adverse affect on labor efficiency and the project schedule.

It follows that certain procurement or approval activities may need to be added to the project schedule.

Economies of Scale

The contractor may propose changes to component sizes to take advantage of economies of scale (Principle 1.3). It is important to recognize that such change proposals may result in changes to methods, which can change activity durations, labor effort (and performance), and activity cost. Changes in methods may negate perceived advantages in economies of scale.

The theory behind economies of scale is that larger components have lower unit costs. The overall labor costs are lower with larger components, and the actual number of components installed is fewer. So, the variable cost of installation of larger components is less, even though the fixed installation cost for smaller and larger components may be close to the same.

The economies of scale view may not always be true when construction methods change. To illustrate, precast floor planks were erected on two almost identical projects. On project BP, the planks were about 115 ft^2, and on project G, they were about 32 ft^2. On project G, the crew size was slightly larger. Applying the concept of economies of scale, project BP should have had the better labor productivity on a square footage basis because the variable costs should be less. Yet, the productivity on project G was better by about 25%, even though the same contractor did both projects. On project BP, the crane lifted one plank at a time, whereas on project G the practice was to lift multiple planks at once. However, the real advantage seemed to be that the smaller planks on project G could be moved into final position manually, whereas with the larger planks on project BP, final movements could only be done with the use of the crane.

So why was project BP an inferior performing project? It appears that on project BP, the larger planks were so heavy that they could not be moved into final position without crane assistance. When the crane returned to the ground level for another plank, there was little or nothing for the crew to do, resulting in much idle time. Conversely, on project G, where the planks were smaller, the crew could manually move the planks into final position simultaneously with the crane returning for another lift (Principle 1.4). There was less crew downtime. By lifting three planks at once on project G, the square feet per lift for the two projects was almost equal. But on project G, there was less crew downtime, so Project G showed superior performance. The larger plank size resulted in a change in the installation method that negated economies of scale.

Fig. 8-1 hypothetically shows the economies of scale pattern for the two projects for manual and equipment-assisted installation. When manual movements are used, the unit costs gradually decrease because the planks are increasingly heavier. At some point, crane assistance is needed. A step decrease in relative performance occurs because of the introduction of much more crew idle time. This example reinforces the notion mentioned earlier that a crane is often an inefficient means of lifting. Cranes are a poor choice for many production-oriented operations because they can result in excessive crew idle time.

Eliminating Tasks

Sometimes it is possible to sequence the work in such a way as to eliminate work tasks. Fig. 8-2 shows the excavation plan proposed by the owner of a wastewater plant. The depth of excavation was about 30 ft. The plan contemplated excavation of areas in the order 1–2–3. If this plan were followed, a retaining wall between areas 1 and 3 would have to be constructed with tiebacks extending to the left. When the area 3 excavation was initiated, the tiebacks would have to be removed and new tiebacks installed in the opposite direction. The contractor proposed doing the excavation in

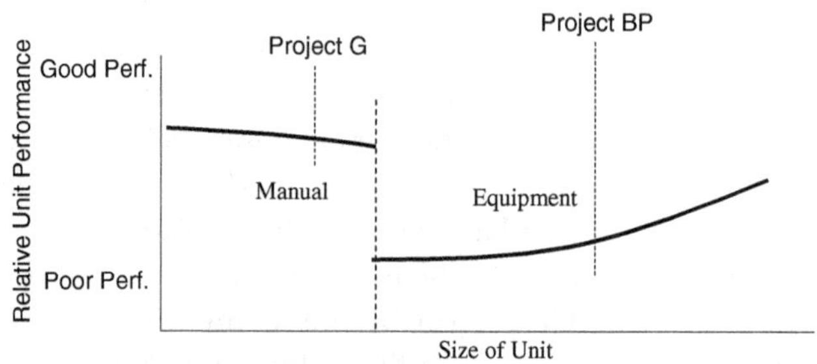

Fig. 8-1. Hypothetical Economies of Scale Pattern for Setting Precast Floor Planks

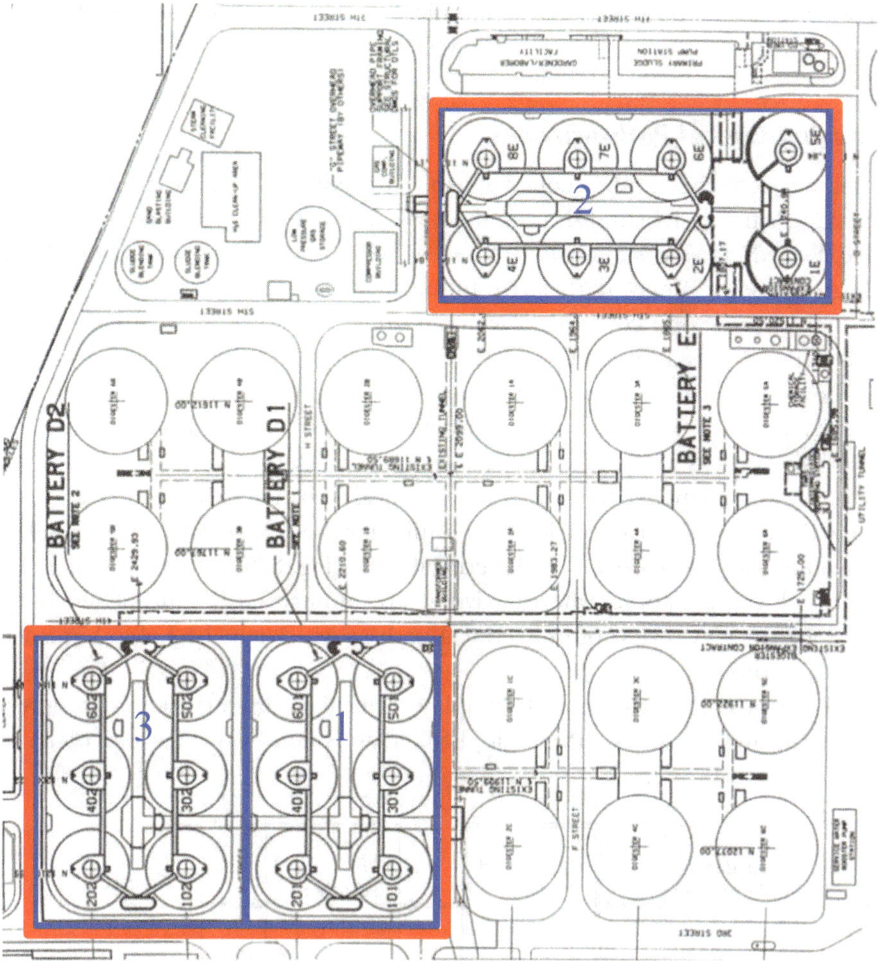

Fig. 8-2. Excavation Sequence for a Wastewater Facility

areas 1 and 3 simultaneously, thereby eliminating the need for the retaining wall and tiebacks (Principle 1.5).

Contractors need to proceed with caution when applying Principle 1.5. If surprises occur, there can be a greater effect on the overall schedule. For instance, if in this illustration, a differing site condition or unexpected groundwater is encountered, both areas 1 and 3 may be adversely affected.

Avoiding Congestion

Avoiding congestion is important to maximizing efficiency (Principle 1.6). Congestion can arise from multiple sources, including multiple trades (stacking of trades), poor scheduling, poor housekeeping, poor material management practices, poor sequencing, and other reasons. Noncritical activities should be scheduled after critical ones.

Managing Critical Path Activities

Sometimes activities can be sequenced so as to reduce the number of critical path activities (Principle 1.7) and shorten the project schedule. This concept is illustrated in Fig. 8-3, which shows a portion of a CPM schedule (four activities) for the State College Municipal Building (SCMB). In this relatively simple example, it is assumed that there is only one concrete placement crew. The work on the basement level (walls and slabs) is all cast-in-place (CIP) concrete. The first floor structural steel must be erected before the first floor precast floor planks can be set. Note that all the activities in Fig. 8-3a are critical path activities. The focus of this example is on the ground slab activity (Fig. 8-4), which does not have to be a critical path activity. All activities that are sandwiched between two critical activities are also critical. However, if the CIP ground slab can be deferred until after the first floor planks are set, then the ground slab is no longer on the critical path (Fig. 8-3b). The scheduler should always seek out situations where activities are unnecessarily sandwiched between critical activities. A review of only critical path activities may be helpful in isolating these situations.

Where possible, one should complete critical path work before progressing to noncritical path work (Principle 1.8). This is one way to avoid congestion.

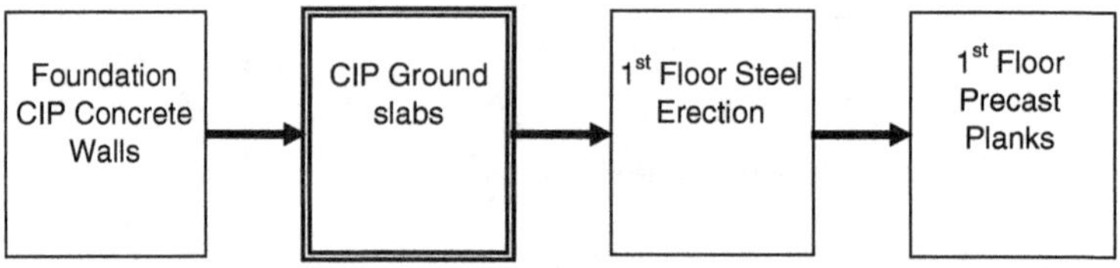

(a) Critical path activities when ground slabs are scheduled early

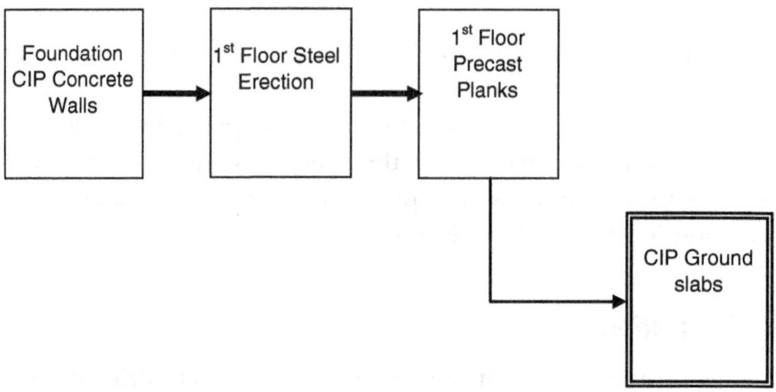

(b) Ground slabs removed from critical path by resequencing

Fig. 8-3. Illustration of Resequencing to Remove an Activity from the Critical Path

Fig. 8-4. Ground Slab that Need Not Be a Critical Path Activity

One should be sensitive to changes in the critical path. Certainly one does not want to be working on noncritical activities only.

Shortening the Project Schedule

Because time is money, contractors should take every opportunity possible to shorten the construction schedule. For example, if a contract allows 300 days, but the contractor can complete the project comfortably in 250 days, it has been observed by the authors that most contractors are likely to take almost 300 days to complete the project. A project shows higher profit if it is completed in 250 days because there is less labor cost and home office overhead. There is also less exposure to risk.

Where congestion or interference can be avoided, work should be done concurrently (Principle 1.9). Doing work concurrently rather than sequentially and taking activities off the critical path can dramatically shorten construction schedules. Changes in construction methods may be necessary, and access can become an issue.

On slow activities, especially critical activities, use multiple workstations (Principle 1.10). Theoretically, using two workstations shortens an activity by a factor of two compared with one workstation. Using multiple workstations requires more tools and materials. Using multiple workstations and scheduling concurrent work instead of sequential work can significantly shorten the project schedule and may offer important labor efficiency advantages. Chapter 3 provided several illustrations.

Material Management

Activities related to material handling may need to be added to the schedule. Preloading is a practice that promotes labor efficiency, but proper sequencing as shown in Chapter 3 is a must. Restocking work areas after regular working hours may be an option to consider.

Utilities and Site Works

Arrange for the early relocation of aboveground utilities (Principle 1.11). Utility relocations may take time. Fig. 8-5 shows an aboveground power line that has not been relocated. This power line was not moved until several months into the project. Belowground utilities should be moved early too. Backfilling should be completed early so that the area can be used for other purposes (Fig. 8-6). Backfilling around basement walls allows easy access into the facility (Principle 1.12). Site drainage facilities should also be done as early as possible (Principle 1.13).

Enclosing and Sealing

Enclosing and sealing a facility is an important milestone. Enclosing and sealing can be done in one step by using permanent windows in lieu of temporary enclosures (Principle 1.14). Enclosing and sealing offers four important advantages: (1) temporary heat can be provided, which can lead to labor efficiencies during winter work; (2) water intrusion can be prevented, which prevents mold; (3) "callback" time can be greatly reduced or eliminated; and (4) labor hours will not be diverted to building temporary facilities. Fig. 8-7 shows a project where permanent windows were installed early. Temporary enclosures may satisfactorily enclose a building, allowing temporary heat, but a temporary enclosure is often unsuitable for sealing a building. Fig. 5-2 shows a typical scenario. Because mold is a serious concern, permanent windows can be a cost-effective alternative.

Fig. 8-5. Aboveground Power Line that Has Not Been Relocated

Fig. 8-6. Late Backfilling that Impedes Access to the Building

Fig. 8-7. Sealing and Enclosing a Building with Permanent Windows

Cost vs. Time Orientation

A building schedule can be somewhat unique in its orientation. It has been observed that some general contractors apply a sequential schedule to the civil phase of the project. This is a cost-oriented schedule. If a project is behind schedule after the civil phase is completed, these general contractors may institute aggressive concurrent schedules of subcontractor work to make up for lost time. This is a time-oriented schedule. In essence, some general contractors rely on others to finish the job on time. A sequential schedule (during the civil phase) is not compatible with a desire to reach the "enclose and seal the building" milestone as quickly as possible.

	Civil Phase	Interior Phase
Primary contractor(s)	General contractor	Specialty contractors
Type of schedule	Largely sequential	Largely concurrent
Orientation	Cost	Time

Buffers (Time Lags)

Buffers are an important aspect of activity sequencing because good productivity is achieved only when a crew can establish and work at its uninhibited pace. Each crew has a different pace, depending upon the work involved. For example, wall formwork is likely slower than footing formwork. Ideally, the fastest activity should be the lead activity in a sequence of activities; the slowest activity should be last. But this is not always possible. When a faster activity immediately follows a slower activity, the productivity of the faster activity will suffer unless there is a buffer or time lag. An analogy is a car on a freeway traveling at 60 mph overtaking another car traveling at 50 mph, except that in the construction analogy, passing is not allowed.

8.2 Case Study 1—State College Municipal Building: Buffers

A number of instances supporting the principles cited in this chapter have been given throughout this text.

A topic important to activity sequencing is the inclusion of buffers (time lags) between activities to allow each crew to work at an uninhibited pace. An uninhibited pace is generally a productive pace. One problem often observed on accelerated projects is that there is such a rush to achieve project completion that adequate buffers are not provided, which leads to idle time and subpar productivity. Loss of productivity is more pronounced where there are considerable differences in the uninhibited daily production rates of the crews involved and a slower crew precedes a faster crew.

This case study involves wall formwork on the State College Municipal Building and illustrates the negative effects of an inadequate buffer. The crews involved are a two-person crew doing trench excavation, a two-person crew installing trench formwork and the key, a concrete placement crew, and a three-person wall formwork erection crew. The complete sequence of activities is illustrated in Fig. 8-8. The completed wall formwork operation can be observed in Fig. 8-4.

Note that wall formwork is last in the sequence of activities. Because it would likely have the slowest uninhibited production rate, there would ordinarily be no problem. However, in this instance, rock had to be removed from about 40–50% percent of the trench. (The western wall and part of the northern wall in Fig. 3-13). The contractor chose to use hand tools to remove the rock. This decision slowed the trench excavation immensely. The wall formwork crew constantly had to wait for the trenching crew to provide adequate work space. This situation continued until the trench crew no longer had to remove rock. Fig. 8-9 shows that the productivity of the wall formwork crew was negatively affected for about 40% of the total time required to complete the work.

Because the wall formwork activity was critical path work, the project schedule was probably extended by a week or more. Also, Fig. 8-9 shows a misleading learning curve effect. Improvement in productivity did not occur because of repetition but because a buffer was eventually created between the trenching and wall formwork activities.

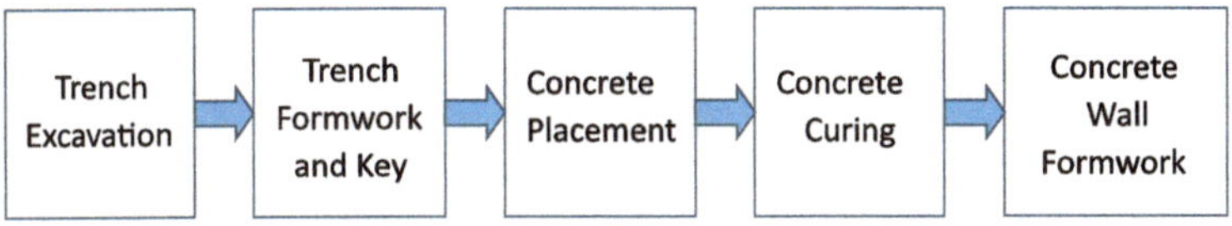

Fig. 8-8. Sequence of Activities for the Construction of Wall Formwork

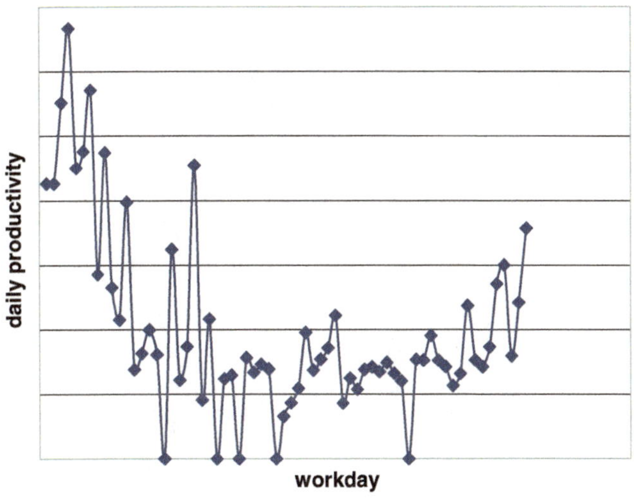

Fig. 8-9. Productivity of Wall Formwork Crew, SCMB

The contractor reacted by working two shifts on wall formwork. The contractor never realized that the root problem was the trenching operation (and the method of rock removal and the lack of a buffer between trenching and wall formwork), not the wall formwork crew.

One effective strategy for this operation is to use two workstations for the trenching crew. The two crews should start at points A and C in Fig. 3-13, thereby saving the area where rock removal was required until last. The wall formwork could be delayed in starting at point C until the trenching crew reached point B. Note that this strategy would also avoid congestion. Other variations to this strategy are possible.

CHAPTER 9

Fundamental Principles of Trade Sequencing

Not all finish and service work is the same. Using a metal stud and drywall system as an example, there are three areas of work that are unique. These are (1) corridors, (2) room and area ceilings, and (3) room and area walls.

9.1 Description of Trade Work

Corridor Work

In a corridor ceiling, there are feeder duct, sprinkler pipe, hot and cold water pipes, feeder conduit and wire, lighting, and the ceiling grid. The work involves four trades or more (carpenters, pipefitters, plumbers, and electricians). The work generally consists of larger components and long straight runs. This work is sometimes called "bulk installation."

Bulk installation has certain characteristics: (1) larger, heavier components and (2) long, straight runs. Bulk installation is usually thought of as occurring in the corridors. There are significant performance implications associated with bulk installation. Work in rooms consists of smaller, shorter pieces. It is harder than corridor work, requiring more work hours per unit of work. Therefore, corridor work must be done efficiently. The actual unit rate of corridor work must be better than the budgeted unit rate for all the work because the harder work still lies ahead. Bulk installation work should probably be at least 20–25% better than the average unit rate used in the bid.

To do bulk installation efficiently, the crew must be properly sized; the corridor should be clear of other crews, stored materials, and trash, thereby allowing for the use of motorized lifts; and the plans should be free of errors and omissions. Adequate hand tools must be readily available. For these reasons, the corridor is usually done first when the corridor is less congested.

The larger, bulky items are installed in the corridor ceiling first, and smaller, field-run items last. The ceiling grid should be done last so as not to interfere with the other installations.

Room and Area Ceilings

The ceiling in the rooms or work areas contains branch duct and diffusers, branch conduit and wire, sprinkler pipe, hot and cold water pipe, lighting, and the ceiling grid. These components are smaller and shorter than similar components in the corridor, and usually they take more work hours per unit to install compared with similar work in the corridor. As in the corridor, lifts, ladders, or scaffolding are required. The ceiling grid and light fixtures are done last because the ceiling grid often establishes the final location of the light fixtures. Figs. 9-1 and 9-2 show situations where the ceiling grid had been installed out of sequence in advance of other work.

Room and Area Walls

The walls in the rooms contain framing, conduit and wiring, receptacles and boxes, insulation, and drywall. In certain locations, hot and cold water pipe and plumbing facilities are present. Painting is also needed. Lifts, ladders, or scaffolding are generally not required. All this work must be carefully orchestrated because there may be as many as six trades involved, and one trade cannot get ahead of another. They must follow a prescribed order.

9.2 Planning and Execution Challenges

As can be seen, the three work zones described here are unique. There are different trades involved, and the methods and the components are different. The prescribed order of construction is unique. Planning for each zone must be uniquely done.

Fig. 9-1. Ceiling Grid Installed before Piping and Conduit Work

Fig. 9-2. Insulation Being Fed through Ceiling Grid

As can be seen in Table 9-1, there are many activities, trades, and subcontractors involved in the services and finishes phase of a building project. There are many opportunities for congested conditions and out-of-sequence work.

Table 9-1. Major Activities and Systems by Zone of Work in a Building Project

Zone of Work	Major Activities and Systems
Corridor	Feeder duct
	Feeder conduit
	Sprinkler pipe
	Plumbing
	Hot and cold water pipe
	Ceiling grid
	Lighting
Room, ceiling	Branch duct
	Branch conduit
	Sprinkler pipe
	Plumbing
	Hot and cold water pipe
	Ceiling grid
	Lighting
Room, wall	Framing
	Electric rough-in
	Plumbing
	Insulation
	Drywall
	Painting

In 1997, research on sequencing and work flow on multistory building construction determined that mechanical and electrical contractors were among the most affected by poor trade sequencing (Riley and Sanvido 1997). Subsequent research showed that interior system contractors (framing, drywall, and ceilings) can also be adversely affected. Table 9-1 includes a list of the major activities in the three primary zones of work in a building.

9.3 Fundamental Principles

Table 9-2 contains fundamental principles of trade sequencing. In planning and executing the service work for efficient operations, the corridor work should be done first (Principle 1.1), followed by the ceiling work in the rooms and work areas. The work on the partition walls should not begin until the ceiling work is complete (Principle 1.2), and the ceiling grid (Principle 1.3) should not begin until after the ceiling work in a particular area is finished.

For there to be efficient construction and timely completion, there needs to be a sequence plan. Every project involving the simultaneous use of different trades and

Table 9-2. Fundamental Principles of Trade Sequencing

No.	Trade Sequencing Principles
1.1	The work in the corridor ceiling (bulk installation) should be done first.
1.2	For maximum labor efficiency, the partition walls in a particular area should not begin until the ceiling work in the area is complete.
1.3	For maximum labor efficiency, the ceiling grid in a particular area should not be started until the ceiling work in that area is complete.
1.4	Develop a sequence plan detailing the order in which the floors or broad areas will be worked and how the work on a single floor or area will be completed.
1.5	Engage all affected subcontractors in developing the sequence plan.
1.6	Begin work in the corridor and then progress to individual rooms.
1.7	Install bulky, heavy items first and smaller field-run items last.
1.8	Allow ample time lags between activities.
1.9	Ensure that there is ample backlog of work so that crews and teams can establish their own pace of work.
1.10	Do not start work in an area unless all the work can be done in one pass through the area.
1.11	Size crews and teams so that different trades work at roughly the same pace.
1.12	Aggressively manage subcontractors.
1.13	Make effective use of subassemblies and modules.
1.14	Plan efficient methods for moving materials to their installation locations.
1.15	Consider making use of the service elevator (which means that it should be installed early).
1.16	Consider preloading materials as the superstructure is being built.

subcontractors or specialty contractors needs to comply with Principle 1.4. The sequence plan needs to include the order in which the floors or broad areas will be done and how the work in each floor or area will be done. The plan need not be complicated or detailed because it is subject to revision. Flexibility is important.

All affected subcontractors should be participants in developing the plan (Principle 1.5). The sequence plan is the subcontractor's plan, and a good plan calls for each party to give up something. If subcontractors are asked to commit to the plan, it is important for it to be their plan. Where one subcontractor takes the lead in developing the plan, there is a high likelihood that the plan will favor that subcontractor. A good plan is one where each subcontractor gives up something for the collective good of the whole. For this reason, the general contractor should facilitate and moderate the meeting(s) where the plan is developed.

The work should begin in the corridor and then progress to individual areas and rooms (Principle 1.6). Corridor work is bulk installation work that is conducive to good productivity because the work consists of long straight runs consisting of larger components. Bulk installation work is important to the profitability of subcontractors because the remaining runs are shorter and are done in more difficult conditions. Therefore, the general contractor should ensure that the corridors remain free from trash and interferences.

Bulky and heavy items should be installed first (at the top of the corridor space), followed by smaller field-run items (Principle 1.7). The work should logically progress downward.

To avoid congestion, there should be ample time lag between subcontractors and trades (Principle 1.8). This method necessitates the partitioning of the floor or broad area into work zones. Time lags can prevent congestion, although some work is symbiotic (see Chapter 7), such as drywall–insulation–rough-in–electric. This work requires close coordination among trades.

The best way for a trade or subcontractor to be efficient is to be able to establish their own work pace. This can only be done with ample time lags (Principle 1.8) and backlogs of work (Principle 1.9). Do not start work in an area unless all the work can be done in one pass through the area (Principle 1.10). Coming back to finish up things is devastating to productivity. The crews and work teams should be sized so that different trades work at roughly the same pace (Principle 1.11).

It is important that the sequence plan be enforced. Out-of-sequence work has a serious negative effect on all subcontractors. Therefore, the general contractor needs to aggressively manage the subcontractors (Principle 1.12). Subcontractors cannot be allowed to schedule their work if it is contrary to the sequence plan. Interior partition walls should not be installed early because they impede the flow of crafts and materials across the floor or area. Fig. 9-3 shows framing that has been installed much too early before the service work has been done. The ceiling grid should not be installed early. Fig. 9-1 shows the ceiling grid installed before some

Fig. 9-3. Partition Walls Installed before Ceiling Work Is Complete

of the service work has been done. If subcontractors are allowed to schedule their work as they please, much out-of-sequence work will be observed.

There can be extensive use made of subassemblies (Principle 1.13) to speed production. Figs. 2-2 and 6-18 show examples of duct subassemblies.

There needs to be an efficient plan to move materials to the work face location (Principle 1.14). A bad situation is for the crafts to carry the materials up the stairs one piece at a time (Fig. 6-19). This practice is a waste of human resources and can lead to accidents. An alternative is to make use of the service elevator for lightweight or bulky materials (Principle 1.15). Of course, this method requires that the elevator be installed early. Consider using a material stocking and resupply crew that works overtime so that materials are available at the start of the morning shift. Preloading practices are beneficial (Principle 1.16) for reducing crew idle time.

9.4 Sequence Plans

For there to be efficient construction, there must be a plan. The sequence plan needs to define (1) the order in which the floors are to be constructed and (2) the order in which the areas on each floor will be completed. Without a sequence plan, chaos may prevail.

The steps in developing a sequence plan are the following:

1. Select the activities and crafts that need coordination.
2. Partition the area or floor into work zones.
3. Determine the starting point and the general direction of progression.
4. Determine the production rate for each crew.

5. Identify the driving crew or activity in the area or on the floor.
6. Develop a schedule based on the work zones.
7. Assign material storage areas.
8. Revise work plans as appropriate.
9. Enforce the plan.

A hypothetical example is given on the State College Municipal Building (SCMB) using selected service activities to illustrate the development of a sequence plan.

1. Select Activities and Crafts

The activities and trades to be coordinated are duct, sprinkler piping, rough-in electric, and drywall.

2. Partition the Area or Floor

The floor partitioning is shown in Fig. 9-4 (the corridor is not shown). In general, each work zone should have roughly the same volume of work. This may not always be possible for every crew. Consider equalizing the workload for the driving crew (step 5). The corridor should be treated as a separate work zone. Seven work zones are identified in Fig. 9-4 (not counting the corridor). The order of the floors is also not shown. The bottom floors are normally reserved for last because they are usually unique and may be the last items fabricated. But if the structure is reinforced concrete, it may be advantageous to the schedule to begin service work on the bottom floor first because constructing the superstructure may be slow and starting at the bottom floor may accelerate the overall schedule.

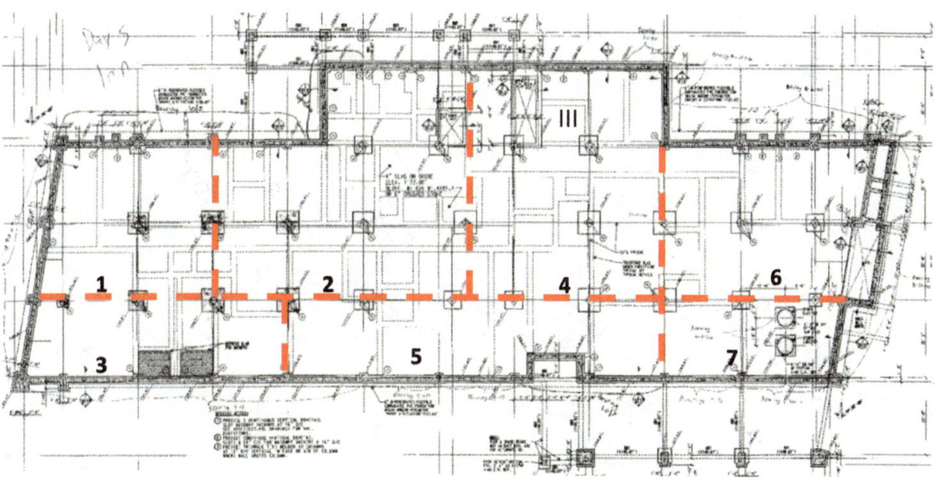

Fig. 9-4. Partitioning into Work Zones of Second Floor of SCMB

3. Determine the Starting Point and Direction

In many instances, the starting point on a given floor or area is immaterial, but give consideration to where services come up from a lower floor. It may be desirable to do the tie-ins first. In this example, the work begins at the upper left of the diagram and generally proceeds to the lower right. The corridor work includes tie-ins with the work on the first floor. The area sequence of construction on the second floor is 1–2–3–4–6–5–7. This sequence allows the north side to proceed in advance of the south side. This scheme may reduce congestion and make allowances for two crews. The sequence of the floors (Fig. 9-4) is 3–2–1–Basement. If the schedule is tight, the sequence may be reversed, thereby accelerating the schedule by maybe two weeks.

4. Determine Production Rates

Hypothetical rates are shown in Fig. 9-5. Ideally, one would like the production rates of all activities to be the same, but this rarely happens. One solution, if the schedule permits, is to allow for a time lag or buffer between activities. This method allows crews to work at their own pace. This is an important key to productive work. Time lags avoid congestion and minimize the likelihood of out-of-sequence work. A time lag is shown in Fig. 9-5 between the start of piping and the start of electrical work.

5. Identify the Driving Crew

In the absence of time lags, typically one crew dictates the pace of the other crews. A series of velocity charts, like Fig. 9-5, are helpful in identifying the driving crew. The driving crew is likely to be the crew with the slowest production rate, r_{dr}. In Fig. 9-5, the piping crew is the driving crew. If the production rate of this crew is increased,

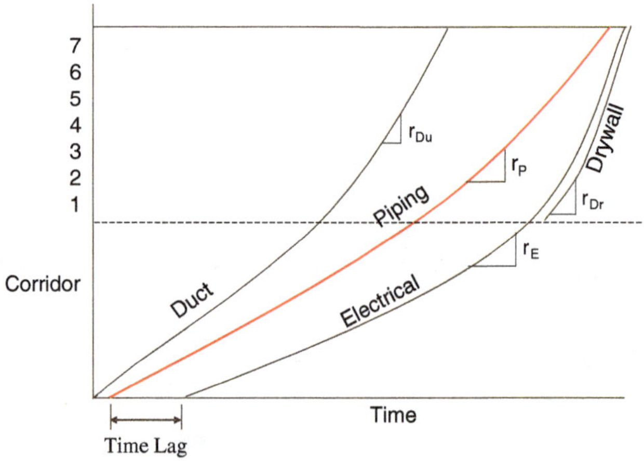

Fig. 9-5. Production Rates, r_i, of Various Activities, Second Floor, SCMB

the floor can be completed in less time. Therefore, one can concentrate improvement efforts on only one crew.

Production rates, r_i, can be changed by

- Increasing or decreasing the crew size,
- Working fewer or more hours,
- Using subassemblies,
- Using multiple crews and workstations, or
- Revising work scope.

6. Develop a Schedule

A work schedule needs to be developed based on work zones. A hypothetical schedule for a single crew on a floor is shown in Fig. 9-6. It shows the area in which the crew is to be working on a given week. As can be seen, the work progresses in an orderly manner. A master schedule for multiple crews like the one shown in Table 9-3 can also be developed. Using this schedule, one can ensure that not too many crews are working on a single floor at once.

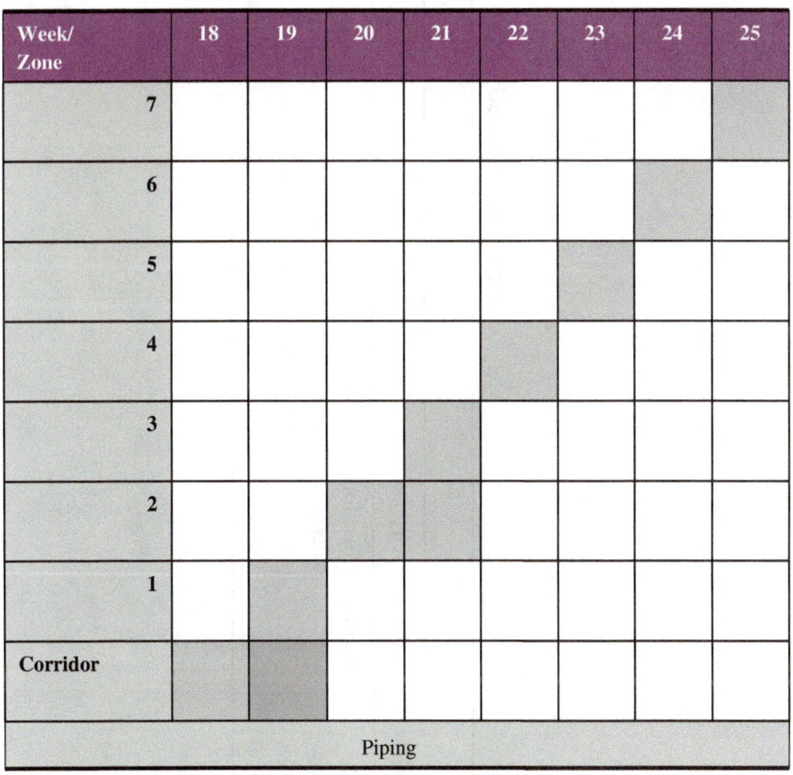

Fig. 9-6. Sample Work Schedule for Piping Crew

Table 9-3. Master Schedule for Multiple Crews

Activity	Week							
	16	17	18	19	20	21	22	23
Duct	Corridor	Corridor	Area 2	Areas 1 and 3	Areas 4 and 6	Areas 5 and 7		
Piping		Corridor	Corridor	Area 2	Areas 1 and 3	Areas 4 and 6	Areas 5 and 7	
Electrical			Corridor	Areas 1 and 2	Areas 3 and 4	Areas 5 and 6	Area 7	
Drywall					Area 2	Areas 1 and 3	Areas 4 and 3	Areas 5 and 6

7. Assign Material Storage Areas

Crews cannot store materials anywhere they choose; otherwise, a situation like the one seen in Fig. 6-9 can materialize. Storage areas should allow each crew access without the movement of materials. A sample storage plan is shown in Fig. 6-27.

8. Revise Plans as Appropriate

No plan is perfect. One should revise plans as appropriate.

9. Enforce the Plan

A plan that is not enforced is useless. Subcontractors cannot be allowed to apply their own plans. It is imperative that subcontractors be involved in developing the plan that is adopted.

9.5 Case Studies

Case Study 5 in Chapter 7 should be reviewed at this point. One can only imagine the chaos that existed on this project as crews searched for a place to work. The out-of-sequence work appears to have been partly or wholly caused by the general contractor's failure to complete the floor slabs in an orderly manner. The out-of-sequence work cost the subcontractor more than $200,000. Effective use of lag times would have minimized this loss.

References

Riley, D. R., and Sanvido, V. E. (1997). "Space planning method for multistory building construction." *J. Constr. Eng. Manage.*, **123**(2), 171–180.

CHAPTER 10

Fundamental Principles for Avoiding Congestion

Congestion is a negative by-product of poor planning. It is often associated with stacking of trades, but it can also be caused by many other circumstances, such as poor material management practices, including material storage and housekeeping; ineffective workforce management practices; out-of-sequence work; and various equipment practices. Scaffolding that is left in place awaiting a crew to return to complete unfinished work can cause congestion. The design can also be a contributing factor when it includes confined spaces. Thus, the principles in Chapters 7–9 should be applied simultaneously with the principles presented in this chapter.

The literature is relatively consistent that workers should have, on average, 250–300 ft^2 per person to work efficiently. Therefore, any situation that reduces the work area available can cause congestion. Most often, congestion occurs on interior work, but congestion can also occur on exterior work.

10.1 Fundamental Principles

With the estimated cost of inefficiencies being significant, many alternative measures are economically justified to avoid congested work areas. Below are measures that can be taken by the general contractor and subcontractors. Table 10-1 lists some fundamental principles to avoid congested work areas. The principles in Table 10-1 should be integrated with the principles given in Chapters 7–9.

Especially for small sites and whenever storage space is limited, the practices of erecting materials directly from the truck and preloading materials (both practices mean that site storage needs are minimal) are highly desirable (Principles 1.1 and 1.2). But when using these practices, delivery schedules must be tightly controlled to ensure that the crew does not run out of materials. It seems easier to control the material delivery schedule if a local supplier (with a shorter travel distance) is used (Principle 1.3). Any practice that reduces the need for material storage helps to alleviate congestion.

Table 10-1. Fundamental Principles for Avoiding Congested Work Areas

No.	Principle
1.1	Erect materials directly from the truck to avoid on-site storage.
1.2	Preload materials on each floor, and store materials to avoid double-handling.
1.3	When applying Principles 1.1 and 1.2, select vendors near the site (minimum travel distance) to better control delivery schedules.
1.4	Where possible, outsource work to others and use preassemblies to move as much work off site as practical.
1.5	Consider putting the project office in a part of the facility being constructed, and avoid moving the office once it has been located.
1.6	When applying Principle 1.5, move the project office into place only after the critical path work in that location is finished.
1.7	Where possible, schedule all noncritical path work at times so as not to interfere with critical path work.
1.8	Maintain good housekeeping.

Outsourcing can be worthwhile to consider. For instance, in masonry operations, mortar mixing on site takes valuable space. Purchasing mortar from a local ready-mix plant (Principle 1.4) can alleviate congested conditions.

On sites of all sizes and configurations, consider building the project office out of lumber and polyethylene on one of the floors of the facility being constructed (Principle 1.5). This facility replaces a trailer, which would take up valuable space. It may also be more readily accessible by the crafts. Construct the office only after critical path work in the area has been completed (Principle 1.6).

A contractor should schedule noncritical path work so as not to interfere with critical path operations (Principle 1.7).

It is important that work areas be free of trash and clutter (Principle 1.8).

10.2 Case Study 1—Bryce Jordan Tower: Disruptions and Congestion

The following case study represents an analysis of reinforced masonry wall construction on a multistory apartment building. The Bryce Jordan Tower is a seven-story apartment building located in State College, Pennsylvania. It consists of an open parking level and six residential floors. The building footprint is rectangular, 58ft 8 in. × 195ft 8in. The footprint area is about 11,500 ft^2, and the site area is approximately 17,500 ft^2. The site layout and building footprint are shown in Fig. 4-2. The slabs on the residential floors are precast concrete. The superstructure, elevated (precast) slabs, and some cast-in-place (CIP) concrete activities were on the critical path.

Project Description

The superstructure is made of reinforced masonry interior and exterior walls. Each cell of the concrete masonry units (CMUs) was reinforced with one #8 bar, and each cell was then grouted. The masonry construction was critical path work.

Precast floor planks were used on all floors to speed construction. The planks were 4 ft wide. The planks were grouted, and a 2-in. leveling course of concrete was placed on top. The setting of planks was a critical path activity. The original design called for welded wire mesh in the 2-in. topping course, but the contractor requested and received permission to substitute fibers in the concrete mix in lieu of the welded wire mesh. This change made the topping operation a simple concrete placement operation.

There were two shear walls on the Bryce Jordan Tower apartment project, one at each end of the building (Fig. 4-2). At one end was a cast-in-place reinforced concrete stairwell, and at the other end was a reinforced masonry elevator shaft.

The original contract schedule was 14 months, with little opportunity for a time extension beyond the end date of July 2003. There was an existing one-story wood-frame dwelling on the site, and the lease lasted through May 2002. The owner chose to take responsibility for the demolition of the structure and removed this work from the construction contract, so the project was one month late in starting. Thus, there was a de facto 13-month schedule. This was an aggressive schedule.

The planned start date was May 1, 2002, but the actual start date was June 3, 2002. The original completion date was June 30, 2003, but the contractor received a one-month time extension until July 30, 2003. The actual completion date was August 8, 2003. The preconstruction period was about two weeks.

The project was expensive, about $99 per square foot. The bid price was $6.8 million. Several factors contributed to the high cost, including the following:

- The site was small and access was restricted.
- The schedule was aggressive.
- The retaining walls, second floor struts, and some other components were heavily reinforced and massive.
- The reinforced masonry walls were labor intensive.
- The Dryvit exterior cladding system was labor intensive and required exterior scaffolding, which was also labor intensive.
- There was a limited number of serious bidders.

Activity Description

The activity observed was the masonry construction of the exterior and interior load-bearing walls on the fifth and sixth floors and a small amount of work on the fourth floor. (Masonry for the elevator shaft was not included in the study.) Each cell of the CMUs was reinforced with one #8 reinforcement rod and then grouted. The block size was 8 in. × 8 in. × 16 in. The CMU vendor was located some 50 miles away. The work was performed by a local masonry subcontractor.

The activity subtasks included placement of CMU block, setting vertical reinforcement, and grouting. Mortar was mixed on each floor, but a local vendor supplied the grout. An accelerator was added to the grout when the truck arrived at the site.

The masonry crew consisted of six to 10 workers, which included the supervisor. The ratio of skilled to unskilled workers ranged from 2:1 to 3:1. The work shift was eight and a half hours per day. The crew was organized into teams. Because one person mixed mortar, the work usually took place at two workstations. One team did the exterior walls, and another did the interior walls. A third, smaller team did the grouting. Generally, the placement of block and grouting occurred concurrently each day, although grouting took place only about half the time, which may have contributed to the inefficiencies in this operation.

The work area was congested with materials, scaffolding, and other items related to the work, as shown in Fig. 10-1. A polyethylene-covered structure was constructed to serve as the project office. This structure can be seen in the center of Fig. 10-1. This project office contributed to the congestion problem. The mortar mixing operation (located behind the project office in Fig. 10-1) also contributed to the congestion. The project office and mortar mixing operation were moved to a higher floor as the floor planks on the next floor were set, thus limiting the work space on every floor. In Fig. 10-1, it also appears that other miscellaneous operations may have occurred simultaneously with the masonry operation. The CIP concrete stairwell can be seen in the lower right-hand corner of Fig. 10-1. Also in Fig. 10-1, it appears that the height of the polyethylene office structure is greater than the floor height. If true, the office structure would have to have been moved before floor planks for the next floor could be set. This situation would have been a major planning mistake.

Fig. 10-1. Second Floor of Bryce Jordan Tower Showing Congested Conditions

Interestingly, if the threshold values of 250 and 300 ft^2 per worker are used as suggested by the literature, the Bryce Jordan Tower could accommodate 35–45 masonry workers per floor before congestion would begin to reduce labor efficiency. At no time were there that many workers assigned to a single floor. Instead, congestion was caused by poor housekeeping, material stockpiles, temporary facilities, and other operations.

Productivity Analysis

Fig. 10-2 shows the daily productivity of the masonry operation. The cumulative productivity was calculated as 0.114 work hours (wh)/ft^2. Several days were noticeably disrupted, in particular, workdays 3, 9, 17, 19, 27, and 29; otherwise, the level of performance is rather consistent. Table 10-2 contains entries from the daily diary. There was no work performed on workday 9. The high numbers in Fig. 10-2 indicate that the work on these days was disrupted.

Calculating Labor Inefficiencies—The work was disrupted by several main factors: congestion, weather (cold temperatures and rain), and workforce management issues. Table 10-3 shows a summary of work hour inefficiencies by cause. There are two productivity values that need to be mentioned here. The first is the calculated actual productivity the contractor could achieve (see Fig. 10-2). This value relies on the actual best productivity on days where there is high production (Thomas and Završki 1999). The calculated best masonry productivity for the Bryce Jordan Tower was calculated as 0.082 wh/ft^2. The second productivity value is the expected or estimated best productivity a contractor should achieve. This

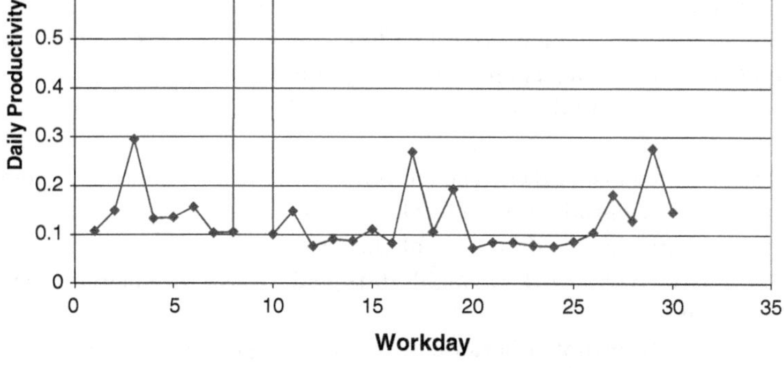

Fig. 10-2. Daily Masonry Productivity, Bryce Jordan Tower

Table 10-2. Daily Diary Field Log

Day	Field Log
1	Construction of the fifth floor began. The masonry crew is constructing the fifth floor walls. Congestion is the main problem affecting performance. A truck-mounted crane lifts the CMUs as they are needed. Light rain is falling, but the work does not seem affected. The crew size is 9 workers.
2	Construction of the fifth floor continues. Light rain continues to fall. Congestion is a growing problem. The crew size is increased to 11 workers.
3	Congestion is affecting the distribution of CMU block to the workstations.
4	Congestion is still a problem.
5	The masonry crew is still on the fifth floor.
6	This day was very cold (<20°F). The crew size is reduced to 10 workers.
7	This was a good day. Masons worked some on fourth floor interior walls, where it was less congested.
8	The crew size was decreased to 8 workers. There was also some rain.
9	No work was done this day. Instead the crew did cleanup in an effort to avoid congestion problems. There was also some rain.
10	The crew size is increased to 10 workers.
11	Crew size reduced to 7 workers because there are few open work areas. This day was very cold.
12	Work was done on interior walls on the fifth floor. The temperature was very cold.
13	Work was done on interior walls on the fifth floor.
14	Work continued on interior walls.
15	Exterior walls on fifth floor finished; team moved to sixth floor.
16	Crew size reduced to 5 workers.
17	Crew size increased to 6 workers. Work was highly disrupted because of cold temperatures that affected grouting. Mortar mixing was delayed five hours.
18	No report.
19	Crew size increased to 8 workers. Cold temperatures disrupted work this day. Mortar mixing was delayed for three hours.
20	No report.
21	Crew size is 7 workers. Crew sizes vary considerably according to the amount of work available and the conditions at the site.
22	Crew sizes vary considerably according to the amount of work available and the conditions at the site.
23	Crew sizes vary considerably according to the amount of work available and the conditions at the site. There was some rain today.
24	Crew size increased to 8 workers.
25	No report.
26	Crew size reduced to 6 workers. The temperature was very cold during the morning hours.
27	Work was highly disrupted because of congestion.
28	Work was highly disrupted because of congestion.
29	The weather was cold. Work was highly disrupted because of congestion. There was also some light snow.
30	Work was highly disrupted because of congestion. The work was greatly affected by congested work areas on each floor and by the small site area, which made crane placement and CMU deliveries challenging.

Table 10-3. Inefficient Work Hour Calculations

				Weather-Related		
Workday	Actual Quantities	Actual Productivity	Congestion	Cold Temperatures	Rain and/or Snow	Workforce Management
1	643	0.107	16			
2	510	0.149	34			
3	276	0.295	59			
4	756	0.133	38			
6	492	0.157		37		
8	583	0.106			14	
9	0					28
11	381	0.148	25			
12	952	0.076		0		
17	270	0.269		50		
19	218	0.193		24		
23	677	0.076			0	
26	468	0.105		11		
27	193	0.182	19			
28	360	0.129	17			
29	161	0.276	13	18		
30	294	0.146	19			
Total			240	140	14	28

value is based on the experience of monitoring and measuring the productivity on numerous projects. The expected productivity for the Bryce Jordan Tower is estimated to be 0.075 wh/ft^2, and it is compared with the calculated best of 0.082 wh/ft^2. As can be seen, this contractor was never able to achieve the productivity that he should have.

The daily inefficient work hours are calculated as the (actual daily productivity minus the estimated or expected best productivity of 0.075 wh/ft^2) × the quantity installed that day. Table 10-4 summarizes the inefficient work hours by cause from Table 10-3. Fig. 10-3 graphically shows the data in Table 10-4.

Effect of Congestion—One indicator of congested work areas on the Bryce Jordan Tower project was the ever-changing crew size. Fig. 10-4 shows the crew size each day. Contractors staff a project with labor consistent with the amount of work to be done. In this instance, congestion and other operations limited the availability of work for the masonry subcontractor, so the subcontractor adjusted the crew size. As seen in Fig. 10-4, the crew size constantly changed, indicating that the conditions and the available work areas were changing daily. The crew size ranged from five to 11.

Table 10-4. Summary of Inefficient Work Hours

Cause	Inefficient Work Hours
Congestion	240
Cold temperatures	140
Rain and snow	14
Workforce management	28
Total	422

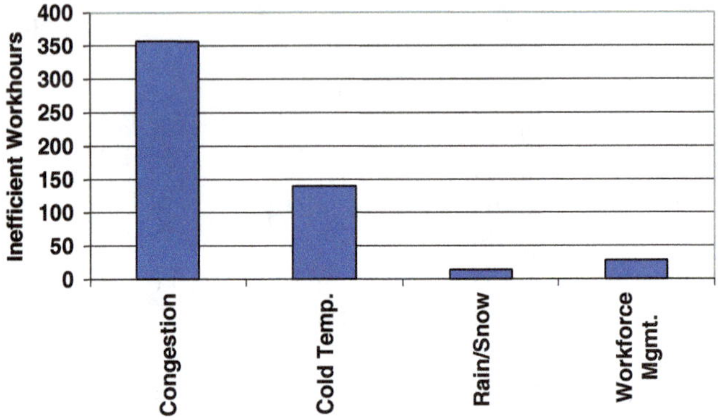

Fig. 10-3. Summary of Inefficient Work Hours, Bryce Jordan Tower

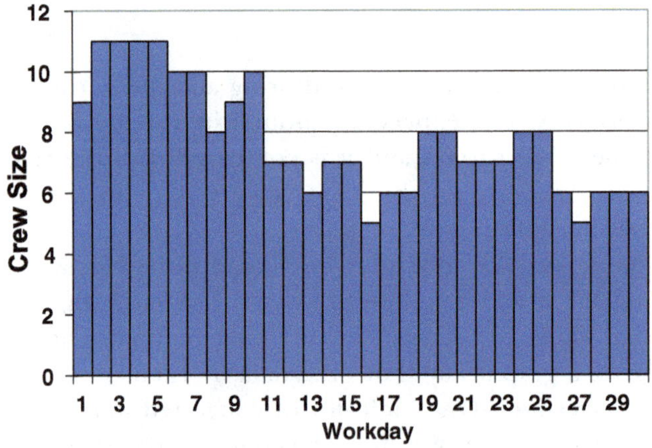

Fig. 10-4. Masonry Crew Size, Bryce Jordan Tower Project

On certain workdays, congestion was noted by the observer as being particularly acute. Notations were made in the diary (see Table 10-2) on workdays 1, 2, 3, 4, 11, 27, 28, 29, and 30. On some days, congestion and cold temperatures were both recorded in the diary, particularly on workday 29. The ambient temperature for these days was noted at 1:00 p.m. to be near 30°F or lower. The principal effect from most of the cold temperatures was that it delayed the start of the mortar mixing operation.

The mortar for the masonry operation was mixed on each floor. This operation required valuable space. It also required work hours to move the equipment from one floor to the next, not to mention the need to move cement, sand, and water to each floor. It seems that the subcontractor would have been justified to use space adjacent to the building footprint to mix mortar and distribute the mortar to the various floors as needed, or purchase mortar from a local ready-mix plant (Principle 1.4). If mortar had been purchased from a vendor, there might have been no cold-temperature delays on workdays 17, 19, and 29. Thus, outsourcing, in this instance, could have been an effective measure to avoid congestion.

The project office structure should not have been moved to a new floor each time more planks were set (Principles 1.5 and 1.6). The move needlessly reduced the space available for the masons. The office could have remained on the second floor. The height of the office structure also should have been made compatible with the distance between floor slabs.

Effect of Weather—Weather effects were of two origins, cold temperatures and rain or snow. Cold temperatures were noted on workdays 6, 12, 17, 19, 26, and, 29. Rain and snow inefficiencies were calculated for workdays 8 and 23. Mortar mixing was delayed for 3–5 hours on workdays 17, 19, and 29 because of cold temperatures.

Effect of Workforce Management—On workday 9, the crew spent the day in cleanup activities (possibly because congestion prevented them from working elsewhere). Because no quantities were installed that day, the inefficient work hours are estimated as the total work hours charged that day.

Inefficiency Summary

The labor inefficiency overall for the fifth and sixth floor masonry on the Bryce Jordan Tower was calculated as 30%, or 422 work hours. The biggest culprit was congestion. Assuming that the work observed covered 25% of the total contract work, the total inefficient work hours for all contract work are estimated as 1,688. At a burdened payroll rate of $35/h, the inefficiencies amount to more than $59,000. The data used for these calculations are shown in Table 10-5.

Table 10-5. Data Used for Inefficiency Cost

Data Category	Fifth and Sixth Floors	Total Project (Estimated)
Total work hours	1,843	7,373
Inefficient work hours	422	1,688
"Should have been" work hours	1,421	5,684
Percentage inefficiency (%)	30	30
Burdened payroll rate	$35	$35
Cost of inefficiencies	$14,770	$59,080

References

Thomas, H. R., and Završki, I. (1999). "Construction baseline productivity: Theory and practice." *J. Constr. Eng. Manage.*, **125**(5), 295–303.

CHAPTER 11

Best Practices for Subcontractor Management

Many types of projects make extensive use of subcontractors. Commercial and high-rise projects may have 40% or more of the contract work done by subcontractors. Residential subdivisions are done almost entirely by subcontractors. Specialty contractors are required by law on public projects in some states. Here, the prime or general contractor may be contractually responsible for coordinating the work of all the specialty contractors employed on the site. It is little wonder that how well the subcontractors are managed dictates, in large degree, the success of these types of projects.

Much has been said about the need to manage subcontractors, but little has been written in the published literature about what specifically a general contractor should do to effectively manage subcontractors and specialty contractors. This chapter summarizes the best practices of industry professionals on how to manage subcontractors and specialty contractors.

11.1 Subcontract Language

Managing a subcontract begins with understanding the language of the contract. Sweet and Schneier provide a good reference text on the legal aspects of contracts (2009). A text by Kelleher offers in-depth exposure to selected aspects of contracts (Kelleher 2005). This text is written by attorneys for engineers and administrators. A text by Thomas and Ellis explains how to interpret contract language. It illustrates how to resolve contract disputes using easy-to-follow flowcharts (2007). This book is written by engineers for engineers and administrators. Nothing comprehensive about subcontracts has been published in ASCE journals.

11.2 Problems with Subcontract Management

Many subcontractors are small, family-owned businesses. Subcontractors may suffer from underfinancing or understaffing, or they may have limited managerial and technical skills (Sweet and Schneier 2009). It is also potentially problematic that a supervisor is often the highest ranking employee on a jobsite. Many supervisors have not had the opportunity to learn expert managerial practices, and this lack of knowledge can negatively affect projects (Sweet and Schneier 2009).

Managing the Work

A major problem in subcontract management is scheduling. Delays occur on many projects. When one subcontractor falls behind schedule, follow-on subcontractors are often delayed in starting their portion of the project. This can create a domino effect, possibly delaying the start of certain subcontractors until after the original date of completion has passed. The reason that delays are such a problem is that "time is money." The longer it takes to finish a project beyond the intended completion date, the more money that must be spent to keep the job running. This results in a negative effect to the general contractor and subcontractors alike.

Schedule delays are not necessarily the result of a subcontractor working at a slower rate than expected. There can be a lack of staff; a lack of equipment, tools, or materials; out-of-sequence work; congestion; changes; unforeseen conditions; or an unrealistic schedule. A lack of staff may indicate that the subcontractor underestimated the amount of work that he or she was required to complete, or the subcontractor may have thought that he or she could work at a faster rate than was possible. Also, the subcontractor may be sharing the labor resources with another project locally that is not yet complete. When dealing with insufficient equipment, tools, and materials, little or no work can be completed. Laborers may stand idle, waiting for the opportunity to work. Repairing or relocating equipment to the jobsite can be time consuming. Materials may require long lead times for their delivery, which can seriously hinder a project. Or maybe the superintendent just didn't order enough materials. Unforeseen conditions can exist that can have significant effects on the schedule, thus necessitating extensive revisions to the plans. Typically, there are a multitude of problems that lead to schedule delays, not just one cause. There can be surprises, too. Fig. 11-1 shows a "surprise" sinkhole in the center of a building on the Penn State University campus. The contractor was paid for this surprise, but the schedule was greatly disrupted.

The lack of a realistic schedule can lead to serious delays and disruptions on a jobsite. When workers are not clearly informed of expectations, the results often do not resemble the original expectation of the general contractor. Schedule changes cause deviations from what was originally planned. A subcontractor termination can also seriously delay the project.

Another problem on the jobsite that arises with subcontractors is that "often, everyone is late." When a subcontractor sees that another subcontractor is not going to finish on time, they may decide that they can take their workers to another job. However, when the time comes for them to return to work on the original project, they are "nowhere to be found." The more work that subcontractors are able to complete,

Fig. 11-1. Sinkhole in the Center of the Katz Building

the more money they make. Therefore, smaller, underfinanced subcontractors cannot afford to have workers on a project if they are not generating revenue.

Schedule delays can result in serious cash flow problems to subcontractors. Because the owner is paying the construction manager or general contractor according to how much work is completed, delays can have a detrimental effect on the financial well-being of subcontractors. Monies that were expected to have been received may not come until weeks or months later. If the general contractor or construction manager does not receive money from the owner, they may not pay their subcontractors. Without income, subcontractors may not have enough funds to purchase materials or pay their workers. Subcontractors may be forced to abandon the job.

Another issue that can arise on a construction project relates to discrepancies in the drawings. If there is a lack of information on the drawings or one drawing shows one thing and another drawing shows something else, delays can occur.

Owners may not freely grant time extensions. The owners may say, "Let's wait until the end of the job and see if you need an extension," or they may wait long periods before granting a time extension, or they may simply say to the general contractor that no time extension will be granted. These situations often create the need for schedule acceleration (this topic is discussed in Chapter 12). Schedule acceleration can be expensive, and some subcontractors may not be able to avoid bankruptcy. Also, some subcontractors may not be able to accommodate the increased staffing demands of schedule acceleration.

Managing People

One problem with managing subcontractors is that of personalities. Some people don't get along with other people. If subcontractors cannot trust one another or management, then the work environment becomes unhealthy. People begin to constantly doubt information coming from other parties, continuing to foster animosity and resentment of others. Regardless of the magnitude of conflict, the inability of subcontractors to work together harmoniously can prevent the project from being completed in a timely manner.

Depending on the subcontractor involved, business understandings may not exist. Subcontractors may not be sophisticated. Some have just gotten their first fax machine only five years ago—forget the Internet. And some still have the owner's spouse doing the accounting books at the kitchen table. The inability to have all parties involved with a project working on the same level can present problems. Communication becomes difficult when e-mail is not being used by all. Also, necessary paperwork may not be received regularly. This lack of understanding of how a business can be run efficiently in today's fast paced and high-demand society may come as a result of many subcontractors being family-owned businesses. After the second or third generation of ownership, the company may not run efficiently. Companies are passed down to family members who may not have the best interests of the company in mind or are incapable of running a business. Regardless of why a company is not succeeding to the best of its ability, a project can suffer as a result of poor business understanding.

Safety is another aspect of a subcontractor's practices that can be subpar. Many construction companies have different views and practices when dealing with the issue of safety. Some subcontractors may have low standards for safety that do not meet the requirements of the owner, construction manager, or general contractor. Without proper safety precautions, workers can be placed in dangerous situations. Fig. 6-10 shows a project where handrails have not been provided around stairs and openings, an OSHA violation.

11.3 Identifying the Controlling Subcontractor

Where there are several subcontractors working, one after the other, it is important to identify the controlling contractor. One subcontractor "drives the train" in that it controls the pace of work of the others. The controlling subcontractor is usually determined by its slow daily production rate compared with the other subcontractors. This subcontractor may need to be given preferential treatment.

Often the controlling subcontractor is the first or lead subcontractor in the string of subcontractors. Velocity charts, like Figs. 3-17 and 3-19, may be useful in making this determination. The controlling subcontractor may not always be the lead subcontractor. Fig. 7-29 is an example where the fireproofing or duct installation subcontractor may be the controlling subcontractor. The important point to remember is that if the controlling subcontractor is delayed or cannot meet the planned schedule, then everyone is affected and the project suffers. If the controlling subcontractor performs as planned, the other subcontractors are likely to fall "in line" and also perform as planned. Thus, the challenge of managing subcontractors may be made easier. Manage the controlling subcontractor, and the need to manage the others may be minimal or at least easier.

11.4 Fundamental Principles

Table 11-1 lists best practices. Practices are listed in two broad areas: managing people and managing the work. These practices support the goal of developing trust and long-term or team relationships with subcontractors. Developing long-term relationships eliminates the need for bid shopping.

Table 11-1. Best Practices of Subcontractor Management

No.	Principle
	Managing People
1.1	Involve all subcontractors in developing the project plans and schedule.
1.2	Build a trust relationship by treating subcontractors fairly.
1.3	Do not engage in the practice of bid shopping.
1.4	Seek commitments from all parties at a prebid meeting.
1.5	Help the subcontractor do timely work by providing assistance and resources as appropriate.
1.6	Walk the job frequently. Get to know the subcontractors' workers and offer assistance as appropriate.
1.7	Explain expectations to subcontractors before bid submission.
1.8	Meet regularly with subcontractor supervisors individually.
	Managing the Work
2.1	Identify the controlling subcontractor.
2.2	Write a fair and balanced subcontract.
2.3	Develop a submittal schedule and change order log.
2.4	Pay subcontractors on time.
2.5	Prequalify subcontractors based on their previous work, safety, and financial situation.
2.6	Require the subcontractors to hold weekly toolbox meetings.
2.7	At regular intervals, evaluate the subcontractor's performance.
2.8	Require subcontractors to maintain good housekeeping.
2.9	Require subcontractors to maintain safe working practices.
2.10	Consider the development of coordination drawings.
2.11	Enforce the contract.
2.12	Every proposed change order should be reviewed by all subcontractors.
2.13	Meet regularly with subcontractors collectively.

Managing People

Principle 1.1. Involve All Subcontractors in Developing the Project Plans and Schedules—A subcontractor cannot be expected to be fully committed to a schedule unless he or she participated in its development; ownership of the schedule is important in developing trust. The schedule also communicates the plan (see Chapter 3). The general contractor should monitor the schedule of the controlling subcontractor, and maybe others, frequently and actively teach subcontractors the best practices of management. Monitoring progress is much easier when the area is partitioned into subareas or work zones.

Principle 1.2. Build a Trust Relationship by Treating Subcontractors Fairly—Trust is built by being open, honest, and fair. A subcontractor wants the general contractor to live up to any prebid commitments that were made. Subcontractors want to be treated as important parts of the project team.

Principle 1.3. Do Not Engage in the Practice of Bid Shopping—Bid shopping is viewed by some as unethical. Relationships of trust are never built by bid shopping. Arditi and Chotibhongs suggest that bid shopping often leads to the selection of unqualified subcontractors who entered the bid process late and may not have adequately planned for the job (2005). Other consequences of bid shopping can be lower standards of work performance, reduced overall project quality, adversarial relationships, legal disputes, unfair competition, insolvencies, and reduced safety performance (Arditi and Chotibhongs 2005). Owners should require general contractors to list their subcontractors and their quotes at the time of the bid.

General contractors should be concerned about the profitability of subcontractors. A profitable subcontractor is a happy subcontractor, or at least content. Being concerned is an integral part of building a team. A "hands-off the subcontractor" policy is probably not a desirable one.

Principle 1.4. Seek Commitments from All Parties at a Prebid Meeting—Conduct a prebid meeting and require all subcontractors to attend. Explain the overall strategy in completing the project and commitments to the schedule. Explain also how the general contractor will handle submittals, payment requests, and enforcement of the subcontract; how the service materials will be brought into the facility; material deliveries; and how and when you plan to enclose the building. This team meeting will go a long way toward building trust. Obtain commitments from all parties relative to the schedule and safety. The general contractor should also make commitments. The prebid meeting should be mandatory and held before subcontractor quotes are developed.

Principle 1.5. Help the Subcontractors Do Timely Work by Providing Assistance and Resources as Appropriate—Subcontractors often do not have in-depth managerial or technical expertise, adequate equipment, and other resources needed for timely completion. The general contractor should consider extending a helping hand, especially to the controlling subcontractor, and maybe others, beyond the commitments made in the subcontract.

Principle 1.6. Walk the Job Frequently. Get to Know the Subcontractor's Workers and Offer Assistance as Appropriate—The opportunity to build trust while confined to a job trailer is limited. Walk the job often and get to know the subcontractor's management team and workers. Offer managerial, paperwork, and technical advice or assistance to the controlling subcontractor's senior representative as appropriate.

Principle 1.7. Explain Expectations to Subcontractors before Bid Submission—Expectations that need to be explained include subcontractor planning and scheduling, production rates or staffing levels, quality, safety, and perhaps other aspects of the job. Some of this can be done at the prebid meeting. Detail expectations and obtain commitments in time for subcontractors to include the effect of these expectations in their bids.

Principle 1.8. Meet Regularly with Subcontractor Supervisors Individually—It is important to meet with the subcontractor's supervisor or senior representative individually and privately on a regular basis to ascertain his or her views on the project, relations with other subcontractors, and factors inhibiting progress. The meeting can be as simple as an informal one at the work face. Regular meetings can go a long way to establishing trust and a line of communication. This is a good time to listen instead of talk.

Managing the Work

Principle 2.1. Identify Controlling Subcontractor(s)—A controlling subcontractor is one that controls the pace of the work of others. There may be more than one subcontractor. If the controlling subcontractor falls behind schedule, other subcontractors will be delayed, too. Therefore, it is imperative that the general contractor do all that is possible from letting this happen.

The general contractor may need to give the controlling subcontractor deferential treatment in the form of managerial, technical, document, and planning assistance beyond the requirements of the contract. Imagine what will happen if the controlling subcontractor falls behind schedule; other subcontractors will fall behind too, and some subcontractors may leave the job. A mess will be created that litigation will not fix. It is useless to engage in finger pointing long after the project is finished. Everyone may be exposed to liquidated damages.

Principle 2.2. Write a Fair and Balanced Subcontract—The authors have seen many subcontracts that are harsh and not balanced. The main purpose of many of these contracts is to avoid payment, take advantage of the subcontractor, or shift all risks to the subcontractors.

Relationships built on trust are not created by contracts that attempt to shift all the risk to the subcontractors. Clauses that are particularly relevant in subcontracts only are the pass-through clause and the "pay if paid" clause. The termination clause and a no-damages-for-delay clause are also important, as may be others.

Principle 2.3. Develop a Submittal Schedule and Change Order Log—A submittal schedule should be developed that has required due dates and actual dates of submission. It is not uncommon for work to be delayed because submittals were submitted late. There should be strict compliance with due dates. Change order and logs of requests for information should also be maintained.

Principle 2.4. Pay Subcontractors on Time—The cash reserves of most subcontractors are much less than those of general contractors. Therefore, it is imperative that subcontractors be paid on time to buy materials and pay workers. The worst thing to happen is for a lien to be attached to a project. Arditi and Chotibhongs (2005) show that subcontractors often were paid more than 60 days after the work was completed. Based on a survey distributed to owners, general contractors, and subcontractors, all categories of respondents felt that the payment process needed to be improved. Arditi and Chotibhongs (2005) recommend that (1) general contractors should be paid promptly by owners, (2) subcontractors should not work for general contractors that have a reputation for paying late, and (3) subcontractors should negotiate the payment terms before entering into an agreement. Arditi and Chotibhongs also discuss issues related to retainage.

Principle 2.5. Prequalify Subcontractors—Subcontractors should be prequalified on the basis of performance on previous projects, safety, and financial health. As lasting relationships are built, the prequalification process becomes easier. Financial status is important as the general contractor doesn't want a subcontractor to go bankrupt. Bankruptcy can happen in six months or less; all it takes is one bad project. An alternative is for the general contractor to select a reliable subcontractor with whom he or she wishes to team and to solicit only one quote. With this alternative, there is little need for bid shopping.

Principle 2.6. Require the Subcontractors to Hold Weekly Toolbox Meetings—A safe job is important to all. A safe job is also a productive one. Subcontractors should conduct weekly toolbox meetings to discuss safety issues. Topics should include potential safety hazards, unsafe practices, and applicable OSHA regulations.

Principle 2.7. At Regular Intervals, Evaluate the Subcontractor's Performance—The general contractor's senior representative or project manager should meet informally and regularly with each subcontractor's senior representative to evaluate the subcontractor's performance and cooperation and their reaction to the general contractor's directions and suggestions.

Principle 2.8. Require Subcontractors to Maintain Good Housekeeping—Subcontractors should be required to maintain good housekeeping practices in their work areas. Fig. 6-22 shows a situation that is intolerable. Figs. 11-2, 11-3, 6-20, and 6-21 show messy work face areas cluttered with duct, piping, and trash. Good housekeeping is essential to ensuring a safe and productive job, and subcontractors should develop a habit of cleaning the site as they work.

Principle 2.9. Require Subcontractors to Maintain Safe Working Practices—Fig. 11-4 shows unsafe rigging practices by a steel erection subcontractor. This site was clean and orderly, yet this unsafe practice persisted throughout the job. The

Fig. 11-2. Untidy Work Face Area

Fig. 11-3. Sloppy Housekeeping at the Work Face

general contractor's project manager needs to be vigilant. This unsafe practice cannot be observed from the office; the project manager needs to walk the job frequently.

Principle 2.10. Consider the Development of Coordination Drawings—In a commercial facility, condominium or apartment building, or high-rise building, there needs to be a plan for how to proceed from floor to floor and how to complete a single floor. Coordination drawings communicate this plan. A preliminary plan should be available before the bids are submitted. All subcontractors should participate in the development of coordination drawings.

Fig. 11-4. Subcontractor Engaging in an Unsafe Rigging Practice

Principle 2.11. Enforce the Contract—At first, this may seem like an unfriendly or burdensome practice. But eventually, everyone warms to the notion that to understand what they must do, all the subcontractor has to do is to read the contract. Sloppy contract administration can only lead to trouble.

Principle 2.12. Every Proposed Change Order Should Be Reviewed by All Subcontractors—A change order request often affects multiple subcontractors. All subcontractor change order requests should be submitted to the owner as a package and should include a general contractor's request for a time extension (if appropriate). No tardy requests should be accepted.

Principle 2.13. Meet Regularly with Subcontractors Collectively—These meetings should be held frequently, say weekly, and are for the purpose of discussing the progress on the job. Problem solving may be postponed until later. Subcontractor attendees should arrive at the jobsite no later than an hour before the meeting to review their activities so that when someone asks a question, they have an answer.

11.5 Case Studies

The authors have no specific knowledge of good or poor subcontractor management practices, except those practices that are readily observable during site visits. For example, the prime contractor is often responsible for housekeeping. If the prime contractor does a poor job at housekeeping, the subcontractor may have to work in deplorable conditions, even though the subcontractor is not responsible. Many of the case studies in this book illustrate situations where the prime contractor–subcontractor interface was poorly done.

References

Arditi, D., and Chotibhongs, R. (2005). "Issues in subcontracting practice." *J. Constr. Eng. Manage.*, **131**(8), 866–876.

Kelleher, T. J. ed. (2005). *Smith, Currie & Hancock's common sense construction law a practical guide for the construction professional*, 3rd Ed., Wiley, Hoboken, NJ.

Sweet, J., and Schneier, M. M. (2009). *Legal aspects of architecture, engineering and the construction process*, 8th Ed., Cengage Learning, Stamford, CT.

Thomas, H. R., and Ellis, R. D., Jr. (2007). *Interpreting construction contracts: Fundamental principles for contractors, project managers, and contract administrators*, ASCE, Reston, VA.

PART IV

Special Topics

CHAPTER 12

Fundamental Principles for Managing Schedule Acceleration

Schedule acceleration or compression occurs when more work is added to a construction contract and no time extension is granted, or when the completion time is shortened without reducing the scope of work. In reality, schedule acceleration can be a combination of both.

Schedule acceleration means that the work must be done in a shorter time frame, and this shortening usually means an increase in work hours. The contractor has three options to increase the number of work hours: (1) work scheduled overtime, (2) hire more workers (overstaffing), or (3) work multiple shifts. Each option can easily bring negative consequences in the form of labor inefficiencies and greatly increased cost.

Schedule acceleration is expensive because the additional costs of labor add up fast. It is not uncommon to see a work hour overrun during acceleration of 25–50% or more compared with the estimate. At current burdened payroll rates, acceleration on a small job can amount to a cost overrun of $250,000–$500,000. Generally, many contractors do not have the cash reserves to sustain these economic losses. Design firms have smaller cash reserves and may struggle to maintain pace with the contractor. It is little wonder that many accelerated projects end in litigation. The only truism about acceleration is that it costs money, usually a lot. The authors have seen acceleration force contractors into bankruptcy, regardless of the size of contractor.

12.1 Management of All the Resources

One of the secrets to managing acceleration is to recognize that acceleration is often a resource allocation problem. One can easily add, say, 50% more work hours by going from a 5-8 (five days of eight hours) schedule to a 6-10 schedule. But is that sufficient? Hardly. Recognize that all the needed resources will be consumed at a faster rate. Materials will be consumed faster, and there will be more forward purchases; there will be more requests for information (RFIs) and changes per unit of time; tools and equipment will experience greater usage; and the demands for support services, like scaffolding, will increase. One must manage all the resources, not just labor.

Acceleration also affects suppliers and vendors. For instance, if the concrete supplier cannot deliver concrete at a faster pace, you may need to engage a second supplier. The fabrication shop may experience more fabrication errors as the pace quickens. As more RFIs and changes are generated per unit of time, more contractor nonmanual workers may be needed for processing these requests. The design firm may not be able to keep pace, meaning that the contract documents may contain more errors and omissions, and drawings may be issued late. Approvals may also be late, and the processing of shop drawings may be untimely.

Schedule acceleration is a uniquely distressful situation, and some management practices that may be applied on a nonaccelerated project do not necessarily work on an accelerated one. Ordinarily, erection directly from the delivery truck is a good just-in-time material management practice, but this practice may not work satisfactorily on an accelerated project. The purpose of this chapter is to explain the conditions on an accelerated project and offer suggestions to help a contractor mitigate some of the acceleration damages.

12.2 Anatomy of Schedule Acceleration

The following describes an actual project to help one understand how schedule acceleration happened on this project. In this discussion, the subcontractor's work, Two Turners' Electric, was accelerated. The facts are based on the court case of *Gilbane Building Company v. Two Turners' Electric Company* (2007).

> In December 1997 Clear Creek Independent School District (CCISD) solicited bids to build two schools: Space Center Intermediate and League City Intermediate. Each facility covered almost 170,000 square feet. Both schools were large, nearly identical, projects. CCISD hired Gilbane as the general contractor in charge of the construction of both schools. While Gilbane was a national building contractor, it was new to Texas school construction, having previously completed only two or three school projects in Texas. For the construction of these schools, Gilbane hired subcontractors to handle certain aspects of the construction. CCISD imposed an absolute deadline of July 1, 1999 for the completion of both schools.
>
> Based on that deadline, Gilbane, as the general contractor, established the schedule, which was based on no scheduled overtime. Gilbane then solicited bids from subcontractors based on a normal work week and 18 months to complete performance.
>
> Daryl Turner served as the president of Two Turners' Electric Co. (Turner). Mr. Turner was a master electrician who has been in business since 1968. Mr. Turner had extensive experience building Texas schools, as he has previously built approximately 25 new schools of approximately this size, as well as numerous other smaller school projects. As a result of this experience, CCISD prequalified Turner as one of 11 electrical subcontractors certified

to bid on CCISD construction projects. Turner submitted a bid for both schools. Mr. Turner prepared these bids based on the knowledge and experience he gained as a master electrician working on similar projects, the schedule prepared by Gilbane for the construction, Gilbane's representations that normal construction practices would be used, and the representation that the subcontractors would have 18 months to complete the construction of the schools.

According to Mr. Turner, "normal construction practices" for electrical work of this character means the following:

1. The presence of structural steel before the electricians install electrical conduit and wiring, creating a place for the electricians to attach the conduit and wiring;
2. The absence of interior walls when the electricians were installing electrical conduit and wiring, allowing the electricians to use motorized scaffolding in straight runs across the building, thus minimizing the time and number of electricians required to perform this task;
3. Logical scheduling, calling for construction from one end of the building to the other, thereby minimizing repetitive setup and maximizing completion of an area before moving to another;
4. Proper scheduling of the work within the available 18 months, allowing Turner Electric to (a) use its own employees as opposed to more expensive and inefficient contract labor and (b) not use scheduled overtime.

Gilbane awarded the electrical subcontract for both schools to Turner. The two lump sum subcontracts contained nearly identical contract language. Turner contracted to perform all electrical work on both schools. Despite the fact that both subcontracts called for Turner to perform all electrical work for a lump sum price, both subcontracts, as well as the prime contract between CCISD and Gilbane, envisioned that there could be changes in the work and included provisions on how changes would be compensated.

Serious issues arose before and during the early stages of the projects that dramatically affected the construction schedule of both schools. The most serious issue arose out of dimensional errors in the design drawings for the schools. As a result of these dimensional errors, the structural steel was incorrectly fabricated. Correcting these errors turned out to be a complicated and time-consuming process. The problem was so significant that, as of May 6, 1998, barely a week after Turner had executed the two subcontracts for the schools, Baker Concrete, the foundation subcontractor, wrote Gilbane saying "... the dimensional mistakes ... have been known for well over three weeks. To date, we still have no complete answers. This was not intended to be a 'Fast Track' project, but at this point it should be treated as such by the design team to maintain progress."

Two days later, Gilbane wrote a letter to the architect stating that the dimensional issue "... turned out to be complicated and unavoidably disruptive bringing about a significant delay to the fabrication and delivery of structural steel as a direct result of these dimensional problems." Gilbane notified the architect that it was completely suspending work in the affected areas of the projects and told the architect "we are unable to continue and we are being constructively delayed in our efforts with resulting impacts to project time and costs."

While Gilbane was not responsible for the dimension errors and the initial delays caused by the errors, Gilbane made decisions in its handling of the dimension problem that directly and adversely affected Turner's performance. Each of these decisions resulted in Turner's work being compressed into a time frame significantly shorter than the 18 months originally envisioned for the work.

Initially, Gilbane made the decision to give priority for the fabrication of structural steel to a Houston Independent School District ("HISD") school Gilbane was building, even though the two CCISD schools had already been delayed while the dimensional issue was worked out by the architect and engineers. This resulted in even more delay in the fabrication, delivery, and erection of the structural steel for the CCISC projects. Under the original construction schedule established by Gilbane, the erection of the structural steel for the schools was to have started in June 1998 and was to have been completely installed by the middle of October 1998. Instead, as a result of the delays in the fabrication of the structural steel, as of September 1998, there was no structural steel on site at either school. By March 1999, only four months before the schools were to be completed, there was still structural steel being erected at both schools. This delay directly affected Turner as much of its work, primarily the installation of conduit and wire, was to be attached to the structural steel and therefore could not be performed before the steel itself was installed. The end result of Gilbane's decision to give the HISD school fabrication priority over the CCISD schools was the further compression of Turner's work into a period of time significantly shorter than the originally planned time.

The next decision to directly affect Turner was Gilbane's unilateral decision to put up the interior masonry walls before the structural steel was in place in an effort to make up some of the lost time on the project. According to Gilbane's senior project manager, Gilbane started looking for something it [Gilbane] could do that was productive and at the same time was overall positive to the schedule. Gilbane made this decision without consulting the subcontractors, despite the fact that this action impeded some subcontractors in the performance of their work. Gilbane persisted in following through with the decision, even though everybody questioned the wisdom of doing so. As a result of Gilbane's decision, the interior masonry walls were built to a height of approximately 7 ft. This directly interfered with Turner's performance of its work as there was no longer an open runway the length of the building for the electricians to use motorized lifts to install the long runs of conduit and wire. Instead, the electricians had to resort to the time-consuming and labor-intensive process of using ladders, with the necessity of setting up and breaking down their equipment many times as they moved from room to room. This not only slowed the pace of the electricians' work, but also increased the number of electricians required to perform the work.

Normal construction practice is to begin erecting the structural steel at one end of the building and progress through the building to the other end. Crafts, such as the electricians and plumbers, which use the structural steel for portions of their work, then follow along behind the structural steel installation and, only when their work is complete are the interior walls built. This procedure allows multiple crafts to work at the same time without impeding each other's work because electricians and plumbers are working in different areas of the building.

The third Gilbane decision to directly affect Turner's performance was Gilbane's decision to abandon the original sequence of constructing the schools. Originally, Gilbane intended to follow the "normal" method for constructing a large school building. Under the "normal" method, work starts at one end of the building and progresses through to the other end. This method creates an efficient flow through the project as it allows each subcontractor to work in a single area until it is complete. This increases efficiency as it minimizes repetitive setting up and breaking down of equipment as workers do not have to move back and forth from one area of the building to another, as well as minimizing the amount of interference and congestion caused by too many subcontractors working in a single area at the same time. In an effort to do something other than "sit on our hands," Gilbane decided to require the subcontractors to work in any area of the schools ready for work. Gilbane decided to work on both ends of the school buildings and progress toward the middle.

Turner was adversely affected in several ways by this Gilbane decision. First, the pace of the work slowed as the electricians had to move from one area to another and then back again. Second, because there were more subcontractors working in a single area, efficiency decreased as they got in each other's way. Finally, by leaving the middle of each school until the end (the middle section contained the bulk of the main electrical equipment), Turner's time to complete its work was dramatically compressed as the electricians had to wait until the later stages of the project to install the main electrical equipment.

Gilbane made a fourth decision affecting Turner Electric. Grinnell Fire Systems (Grinnell) was Gilbane's subcontractor for the installation of the fire alarm systems at both schools. In addition, Grinnell sub-subcontracted with Turner for the installation of the data, sound, and telephone systems, but not the fire alarm system. As a result of Gilbane's earlier decisions, Grinnell's work on all of the systems had been compressed until the final months of the projects. As the July 1 deadline approached, Gilbane decided that the fire alarm systems were more important than the data, sound, and telephone systems. In Gilbane's view, if each school had a functional fire alarm system, it increased the chance that CCISD would accept the schools as complete even though significant work remained. Therefore, Gilbane ordered Grinnell to devote all its attention and resources to the fire alarm systems and let the systems Turner Electric was responsible for—data, sound, and telephone—wait. This decision further compressed the time available for Turner to complete its work.

As a result of these decisions, Turner's time to perform the work was significantly compressed. This, in turn, required Turner to institute scheduled overtime and hire more workers, increasing the subcontractor's labor costs. In addition, as the number of workers goes up, efficiency goes down because it is more difficult to supervise a larger workforce. Finally, the compressed time required more subcontractors to work in the same areas at the same time in an effort to get the work done, with a resulting effect on efficiency. Congestion further impeded progress.

Faced with a significantly compressed schedule, Turner worked hard to complete the schools. According to CCISD, the two buildings were a beehive of activity, even late at night and on weekends, as electricians worked to get the buildings completed on time. In spite of all the problems, construction on the two schools was sufficiently complete to allow the schools to open to students on time or only three days late. However, there was still considerable work to be done, even after the schools were opened, as there were entire areas of the schools that had not yet been closed in. This work had to be completed in the evenings and on weekends, times that required the payment of overtime wages to the workers.

As the time approached to close out the two projects, Turner submitted a claim to Gilbane to be paid for the additional labor costs it incurred as a result of the compressed schedule. At this point, Gilbane made the final decision affecting Turner, deciding not to present Turner's claim to CCISD for consideration.

This narrative highlights several points that are common among many accelerated projects. These are the following:

- Many of the events leading to acceleration are beyond the contractor's control, and therefore, a prebid risk assessment is important (see Chapter 3).
- Contractors and subcontractors are left to react to events that occur rather than shape or control the events.
- Many projects are disrupted by structural steel erection, whether it be from fabrication errors, late submittals, poor contractor performance, or other causes.
- Decisions relative to the schedule are made that often exacerbate the original acceleration.
- On accelerated projects, the sequence is often disrupted or abandoned.

In addition, there can be other unintended consequences. Some of these are the following:

- There may be an increase in theft.
- There can be much incomplete work as supervisors and workers may feel that finishing 95% of the work is satisfactory.
- Return visits to work areas to finish the work are not uncommon.
- Work locations can change frequently, leaving supervisors little time to plan their work.
- There may be a large number of changes to the plans resulting from errors and omissions.
- Multiple and lengthy punch lists often accompany acceleration.

12.3 Can You See Acceleration Coming?

It would be advantageous to know that acceleration is looming in the future so that there can be some advance planning to lessen the effect. It turns out that this is a relatively straightforward process in most instances. Fig. 12-1 is used to illustrate this point.

Suppose that a job is estimated to require 100,000 work hours. If the labor is underused early in the project and, say, at the 25% complete milestone, only 6,000 work hours have been charged, the labor resource will have to be overused later.

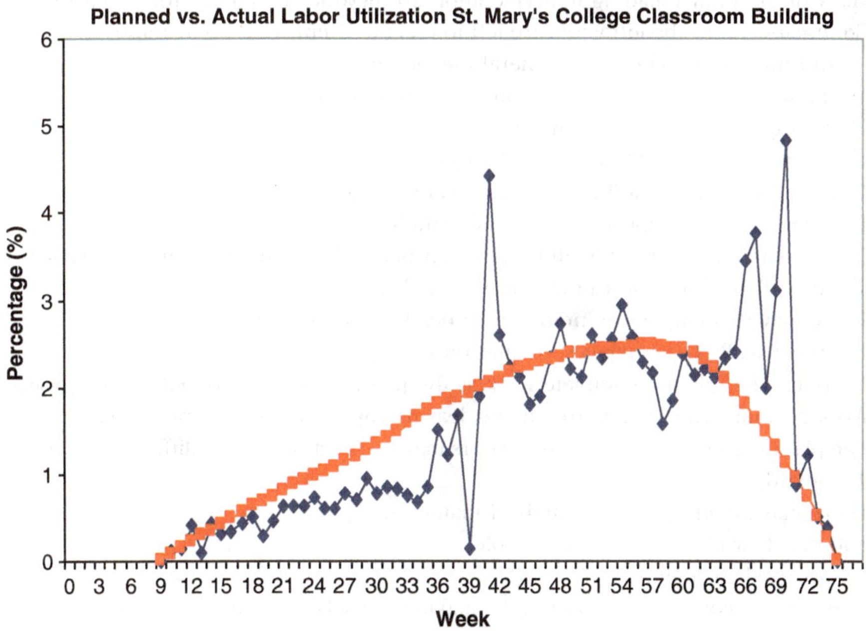

Fig. 12-1. Labor Usage (Planned and Actual) on an Accelerated Project

Fig. 12-1 shows the planned and actual labor usage on an accelerated project. Beginning about week 18, the actual labor usage (expressed as a percentage) is less than planned. This deficit needs to be made up later. About week 40, a large change order was issued that affected the work of multiple subcontractors and specialty contractors. The accelerated work did not begin in earnest until about week 61. The acceleration was predictable, perhaps as early as weeks 20–25. Advanced planning may have mitigated some of the damages.

12.4 The Schedule Acceleration Environment

Planning and Scheduling

To effectively manage schedule acceleration, one must understand how the acceleration environment differs from the environment on an ordinary, unaccelerated project. The accelerated environment is quite different from the norm. Table 12-1

Table 12-1. Partial Listing of Schedule Acceleration Environmental Conditions

Environmental Conditions
• Suppliers and vendors may have difficulty in meeting an accelerated delivery schedule.
• Fabrication shops may experience more fabrication errors.
• Design firms may have trouble keeping up with an accelerated pace.
• Submitted approvals may be late, and the processing of shop drawings may be untimely.
• Many of the events leading to acceleration are beyond the contractor's control.
• Contractors and subcontractors are left to react to events that occur rather than shape or control the events. There is a general loss of control.
• All the needed resources will be consumed at a faster rate.
• There are more forward purchases.
• There can be more RFIs and late changes.
• Tools and equipment will experience greater usage.
• The demands for support services, like scaffolding and lifts, increases.
• Decisions relative to the schedule are often made that exacerbate the original schedule.
• Schedule float time is lost until it is essentially all gone.
• There may be many more intermediate deadlines and milestone dates.
• There is usually much congestion and out-of-sequence work.
• There can be much incomplete work as the pressure mounts to make more progress.
• Work locations can change frequently, leaving supervisors little time to plan.
• The planning horizon for the supervisors may be short. It may be difficult to plan beyond the next day.
• There may be an increase in theft of materials.
• The weekly workload is highly variable.
• A *labor sink* is formed.
• Many sites become messy and untidy, making housekeeping an ever-present concern.
• Time lags dwindle to near zero.
• Multiple and lengthy punch lists often accompany acceleration.

summarizes some of the more common features of the acceleration environment. In accelerated situations, the best laid plans may be rendered useless, and a complete rethinking of the plan is probably justified. Schedule flexibility is important. Many things happen that are beyond the contractor's control. Contractors are often relegated to just reacting to events affecting the schedule, rather than shaping things that happen. Contractors may have little control. Fig. 4-3 shows a situation where the precast erection subcontractor does not have access to the much of the site because of the conditions at the site. Also, the completion of a duct bank passing through the site was delayed and prevented the orderly repositioning of the crane. The only option available to the subcontractor was to go outside the perimeter fence with the crane and progress along the public street.

Because there is so little control, the planning horizon for the supervisors may be short. It may be difficult to plan beyond the next day. Look-ahead schedules may be of little value. Because changes in the work plan may occur daily, the supervisor has great difficulty in planning ahead. Many would agree that planning ahead is perhaps the most important function that the supervisor is asked to perform. Yet, the supervisors who normally do their planning on a weekly basis are stretched to the limit, such that short-range planning is greatly hampered. The turnaround time for taking positive action is relatively short.

As the acceleration deepens, more and more schedule float time is lost, until it is essentially all gone. There may be many more intermediate deadlines and milestone dates. Out-of-sequence work and congestion are common.

Late changes often cause or accompany acceleration. Many times, the need for a change is identified early in the project execution, but the actual change order is not issued until late in the execution time frame. Sometimes, the owner has difficulty deciding what to do. There may be many areas of the project where the work has been put on hold. The demands on the information and material supply networks increase.

A concentrated effort is often made on major work items, leaving other work until later. The last 5% of the work may take months. Multiple and lengthy punch lists may be generated.

Often events occur that form the basis for the owner paying in an untimely manner or not at all. The owner may deduct certain work from the contractor's scope of work, thus further exacerbating the contractor's cash flow and scheduling situation.

The weekly workload is highly variable. One week there may be much work to do, and a worker may work 55–60 hours that week. The next week may find that the available work has dried up, and the same worker may work only 35 hours that week. Thus, one can generally track the amount of work available to do by examining the number of hours worked in a given week by individuals or crews. A graph like the one in Fig. 12-1 is invaluable to understanding issues like workload, and the graph should be produced regularly.

Labor Demands

Two situations commonly occur on an accelerated project. There is a need for additional labor, and there is a highly variable workload. The need for varying amounts of craft labor creates a difficult situation for a superintendent, especially in a tight labor market, because a contractor cannot realistically terminate a sizable part of the workforce during slow periods only to hire them back a short time later when the workload increases. They simply will not be available. What is one to do?

Labor Sink—Often, the excess workers who are not needed because the workload is low are not terminated, but rather are sent to some geographical location on the project where they can congregate. There is limited or no work to do in this area. This "holding" area is called a *labor sink.* It is usually at a remote area of the site, out of view of the superintendent. If visible, the labor sink would be a constant reminder to the superintendent that the project is progressing poorly. The need for a labor sink is one reason (but not the only one) that schedule acceleration is so expensive. If the contractor files a claim, the labor hours in the labor sink should be included. This can be a sizable number.

Variable Workload—A variable workload is a common feature of schedule acceleration. Whether it comes from design changes, unforeseen conditions, material delivery delays, or other causes, there are times when there is little work to do and there are times when there is a lot of work. A variable workload creates a variable demand for labor. This situation can be observed in two ways. First, the crew size may vary appreciably from day to day. Second, the work hours per week may vary. When there is much work to do, a crew may work 50 hours or more during a given week. The next week may see only 35 hours as the available work dwindles. The idea of a crew working a 50-hour work week every week during scheduled overtime for, say, 15 consecutive weeks is incorrect because the workload does not remain constant.

Overtime and Overstaffing—When overtime is necessary, it can be instituted quickly. But fatigue can be problematic. Therefore, overtime is more of a short-term solution to the acceleration problem. If overstaffing is necessary, the newer members of the workforce may be less skilled and will be unfamiliar with the site. They may not work as diligently as those with a longer tenure on the job. Overstaffing is a long-term solution to acceleration.

Other Conditions

Construction errors may be more common. When acceleration occurs, there is much incomplete work. Many sites become messy and untidy. Trade sequencing principles are not followed, and there is much out-of-sequence work. Time lags dwindle to near zero. It seems that every subcontractor is on his or her own. Punch lists are more extensive, and labor needs can be difficult to forecast.

12.5 Fundamental Principles

A list of acceleration principles are described in Table 12-2. Several principles relate to acceleration in general.

Prebid Principles

It is difficult or improbable to make a profit on an accelerated project. More often than not, a contractor loses money. Therefore, a risk assessment of acceleration should be made before deciding to submit a bid. The risk assessment can be tailored to any contractor or any project. Some of the relevant questions to ask are the following: What is the likelihood of a time extension? What is the likelihood of acceleration? Do the plans appear complete and generally free from errors and omissions? Will you receive timely payments? Will you be paid at all? Does the contract contain harsh language? Does the contract include a no-damages-for-delay clause? Many other questions may be relevant. These and other questions should be asked before deciding to submit a bid. A good principle to follow is to leave risky projects to your competitors (Principle 1.1).

You should know or inquire about the reputation of all key parties involved in the project (Principle 1.2). This inquiry should include the owner, designer, construction manager, general contractor, subcontractor, vendors, fabricators, and other parties important to the success of the project. Are they easy or difficult to deal with? Do they engage in the practice of bid shopping? What is the likelihood of numerous changes? Will they pay? Are their management practices sound? Will you be treated fairly? Are the vendors' delivery schedules reliable? Are their products free from defects? Do they stand behind their products? If acceleration occurs, can others keep up with the accelerated pace? Does the owner seem to be well financed?

Preconstruction Principles

One truism about schedule acceleration is that it costs the contractor money, potentially lots of money. However, if the contractor spends money wisely, it is possible to minimize the economic losses. Many of the principles listed in Table 12-2 have cost implications, but they also have the potential for saving significant sums of money. Two of the more costly elements of acceleration are labor inefficiencies and unabsorbed home office overhead. When one considers the burdened payroll rate and the potential for improving labor efficiency, almost any strategy for improving labor efficiency will be cost-effective. Thus Principle 2.1 is predicated on improvements to labor efficiency and states that acceleration is not necessarily the time to tighten the purse strings and not spend monies that may potentially save substantial sums of money.

Table 12-2. Fundamental Principles for Managing Schedule Acceleration

No.	Principle
	Prebid
1.1	Perform a risk assessment of acceleration before deciding to submit a bid and leave risky projects to your competitors.
1.2	Know the reputation of all the parties involved in the project.
	Preconstruction
2.1	While the manager must monitor expenditures carefully, acceleration is not necessarily the time to tighten the purse strings and not spend monies that may save substantial sums of money.
	Construction Execution
3.1	Plan for the accelerated use and demand for all resources, not just labor.
3.2	Improve communications among crews.
3.3	Search for ways to make critical activities noncritical.
3.4	Reduce tasks of large scope into multiple tasks of smaller scope.
3.5	Monitor progress more frequently.
3.6	Set daily or weekly production goals.
3.7	Do more detailed operational planning using short interval production schedules (SIPS).
3.8	Maintain logs of submittals, RFIs, changes, and correspondence and daily diaries.
	Labor
4.1	Consider using a 4-10 or a rolling 4-10 schedule.
4.2	Monitor absenteeism for detrimental trends.
4.3	Add more supervisors as the workforce increases.
4.4	Shift to smaller crews or teams.
4.5	Insist on greater backlog of work before beginning work.
4.6	Stress the crew performing 100% of their work before moving to another task.
4.7	Be selective about the work assigned to a second or third shift.
	Materials
5.1	Assign a material coordinator to the project.
5.2	Use modular or preassembled components.
5.3	Deliver as much material to the project as is practical as quickly as possible.
5.4	Reschedule long lead items.
5.5	Organize a special cleanup crew.
5.6	Be sensitive to theft.
5.7	Increase the inventory of expendables.
	Information
6.1	Add supervision to identify design errors and omissions.
6.2	Organize a special crew to handle changes.
6.3	On every change order, include a disclaimer reserving the right to claim change costs at the conclusion of the job.
6.4	Don't forget to comply with the notice clause in the contract.

Table 12-2. Fundamental Principles for Managing Schedule Acceleration *(Continued)*

No.	Principle
	Tools and Equipment
7.1	Increase the inventory of spare parts and hand tools.
7.2	Sharing of most tools is not a good idea.
	Support Services
8.1	Evaluate the need for more scaffolding.
8.2	Evaluate the need for more lifts.
8.3	Evaluate the need for more lifting capacity.
8.4	Evaluate the need for more formwork.

Construction Execution Principles

When a schedule is accelerated, one usually increases the labor resource via scheduled overtime or by hiring more workers (overstaffing) with the expectation that more craft hours assigned to the work will yield higher outputs. When acceleration happens, the use of and demand for all resources increases. Key resources include labor, materials, tools and equipment, information, and support services, like scaffolding and lifts.

If labor hours are increased by, say, 50%, then the materials consumption rate increases theoretically by 50%. Delivery schedules will become strained, and it may be necessary to acquire some materials from another vendor.

Tools and equipment will be used faster, so expect that you will need more drill bits, grinder wheels, and other materials. Power tools may wear out faster. More hoisting and lifting capacity may be needed. If there are more crews, some tools and equipment are assigned on a per crew basis. The idea of multiple crews sharing certain tools is probably not wise because hoarding can be a constant problem.

Because the work is being installed faster, expect more RFIs and changes per unit of time. It may be necessary to bring on more contractor nonmanual personnel to handle the increased demand.

Support services include scaffolding and quality control. You may need more scaffolding and lifts. You don't want the amount of work you can do to be constrained by the amount of scaffolding you have available. Site conditions must be controlled to ensure that lifts can be used effectively. You may need a lift for each team, and you may need more as the job nears completion. Do not overlook concrete formwork because there may not be sufficient formwork quantity on site for an accelerated schedule.

During acceleration, communication links between crews often become ineffective. Promote good, informal communication at the supervisor level. Allow the supervisors to develop suitable work schedules. Management should intercede only when necessary.

Labor Principles

Consider using a 4-10 schedule (Principle 4.1). Workers like a 4-10 schedule because they get a three-day weekend every week. The long weekends act as a buffer against the fatigue caused by a grueling project schedule. The available research (which is sparse) suggests that a 10-hour workday is not detrimental compared with an eight-hour workday. (A 12-hour workday is probably not a good idea because soon after one returns home, it's time to go back to work.) One advantage of the 4-10 schedule is that there are four starts of the shift and four ends (a total of eight each week). These times are notorious for poor productivity. With a 5-8 schedule, there are 10 such periods of poor performance, an increase of two.

If there is intense pressure to work the project seven days per week, one may consider a rolling 4-10 schedule. Here one workforce works for four days and then has four days off while a second workforce works the job for four days. Thus, the two workforces alternate. A rolling 4-10 schedule is hard on supervision, but it may be a better alternative than a 7-10 schedule. A 7-10 schedule is hard on workers and can probably be worked only for several weeks before fatigue becomes problematic.

A good practice at any time is to closely monitor absenteeism (Principle 4.2). One who is habitually absent is also likely a poor performer, does work of marginal quality, and probably has a poor record of safety. This is the concept of "a bum is a bum is a bum." A good layoff policy is to terminate those with high absenteeism rates first.

As the size of the workforce increases, add more supervisors (Principle 4.3). In general, a smaller crew size is better suited to an accelerated schedule because the negative effects of acceleration are spread over fewer workers and the work near the end of the job is not suited to a large crew. Early in the project, a typical crew size may be six to 10 or more. Later in the project, the crew size may be four to six or fewer. The change in crew size reflects the change in the nature of the work. During acceleration, one should consider smaller crews or divide the crew into flexible-sized work teams (see Chapter 7).

During acceleration, there is intense pressure to complete work; the orientation is toward high production. There is pressure to produce something, no matter how small. Consider the following example. Suppose that there are 14 condominium units on each floor of a 15-story condo project. On one particular floor, the rough-in electric has been completed on a few of the 14 units. The owner pressures the contractor to install whatever insulation can be done so that drywall installation can start. Is this a good idea?

At the outset, recognize that installing insulation is likely to be much faster than the electrical rough-in. The insulation work will "catch up" to the electrical rough-in, and thereafter, the pace of the insulation work will be set by the pace of the electric rough-in crew. For the insulation work to be productive, the insulation crew must establish its own work pace. This will only happen if there is sufficient backlog of work for the insulation crew so that they do not catch up with the electrical rough-in crew (Principle 4.5). There needs to be a time lag between these two crews. This scenario is a common problem in acceleration, that is, time lags tend to disappear. Starting the insulation work too soon is probably not a good idea.

With such intense pressure to produce output, crews sometimes do not complete their work 100% before moving to the next task. The consequence of this situation is that there may be much work left throughout the project that is incomplete. This affects labor in two ways. It is very time consuming and costly to return to a work location multiple times, regardless of how much work is left undone. Second, follow-on crews may need the work to be 100% complete before they can start. For example, pipe must be 100% complete before it can be insulated, and anything less may prevent the insulators from doing their work efficiently.

The likelihood of a lengthy punch list or multiple punch lists is greater. Incomplete work can have a ripple effect throughout the entire project. Thus, project engineers and project managers should stress having the work 100% complete before a crew moves to the next task (Principle 4.6).

There are two strategies that can be applied relative to the work assigned to a second or third shift. First, the later shifts can work on the same task as the first shift and simply advance the progress further. Coordination between shifts with this strategy can be problematic. An alternative is to give the later shifts entirely different work. The first approach is more common, but one problem with the first approach is accountability because problems that arise with the work can always be the fault of the other shift. The preferred approach is to give later shifts different work. The task can be the same, say, formwork, if multiple workstations are used. Fig. 8-4 shows a concrete wall that was partly constructed using a second shift. The second shift was more productive than the first shift by a wide margin. A likely reason is that the first shift advanced the work until it was 90–95% complete. The second shift did a small amount of work to finish the work and received credit for doing 100% of the work. This situation can create animosity among crews.

The type of work assigned to a second or third shift should be carefully considered (Principle 4.7). On later shifts, there is limited design or engineering support available. The site is quieter, and supervision may be sparse. Therefore, select work that will benefit from or will not be hindered by these conditions. Pulling wire probably does not require much input from designers. It also may benefit from a quieter environment because the crew is likely to be spread out. Noisy activities are also good candidates for later shifts. Messy operations, like fireproofing or sanding drywall joints, are good candidates for later shifts. Try to select work that does not require close supervision. For example, if it is expected that a painter can paint two rooms during a single shift, a supervisor can leave the painter unsupervised. The supervisor can return near the end of the shift and readily evaluate if the worker was productive. Knowing how much output to expect is important when using multiple shifts.

Materials Principles

Management must be prepared for accelerated material deliveries, installation rates, and production of waste. Every aspect about materials needs to increase.

Because all aspects of the initial materials plan have to change, it is perhaps helpful to assign a material coordinator to the project (Principle 5.1). This coordinator should probably be a senior person because he or she will have to update delivery schedules, negotiate new procurement agreements, change waste removal plans, revise and enforce the site layout plan, suggest revised construction methods and equipment needs, coordinate with subcontractors, and much more. This is a daunting job, but the success of this coordinator will have much to do with the successful completion of the project.

There can be great efficiency and time advantage to using modules or preassembled components (Principle 5.2). Gang forms and preassembled duct components (see Fig. 2-2) are but two preassemblies seen on nonaccelerated projects. There are many more opportunities.

Because work schedules and plans are constantly changing, it is necessary to have as much material as possible delivered to the site (or nearby) as soon as possible (Principle 5.3). The contractor can ill afford to be unable to proceed to work in newly assigned, unplanned work areas because the materials for the area are not available.

The acceleration may necessitate a new site layout plan that affords more on-site storage. As an alternative to on-site storage, it may be better to rent nearby warehouse space. Of course, plans must be made for daily material deliveries to the site, possibly after normal working hours. Revised methods may be appropriate.

There may be many items with long lead times. For some, the need to renegotiate delivery times is obvious; for others, the need to renegotiate delivery times is not so obvious. Thus, a careful review needs to be undertaken. One example of an item that is important to procure soon is windows. The need to enclose and seal a building is clear. But the facility also needs to be watertight to prevent mold. The lack of "callback" times to replace wet drywall more than offsets any added expense of expedited deliveries.

Not surprisingly, trash is generated faster. It may be justified to have more frequent trash pickups. One effective strategy is to organize a cleanup crew (Principle 5.5). A skeleton crew can keep the site clean during the day, but a full cleanup crew can do a more thorough job as a second shift. While cleanup may seem a trivial task, poor housekeeping has been documented on numerous occasions as detrimental to labor productivity. Many accelerated projects have poor housekeeping practices and tend to become messy and untidy.

It goes without saying that there are many small items that are installed in a facility. Some of these items are not "off the shelf" items and can ill afford to be lost. As the acceleration deepens, there is more opportunity for these items to be stolen or misplaced. One needs to be sensitive to this issue.

The expendables, such as nails and screws, will be consumed at a faster pace. Also spillage and theft may become more problematic. It is probably wise to increase these inventories.

Information, Tools, and Equipment Principles

Late changes can be one cause of acceleration. Even if not a cause, numerous late changes often accompany acceleration. In some circumstances, changes are not an issue at all. But in many acceleration situations, information-related issues are present. Therefore, it is important to have strategies to deal with information overload. Informational issues involve changes, requests for information (RFIs), shop drawings, and submittal approvals. If the work schedule goes from 5-8 to 6-10, then in theory, each of the components of the informational issues will increase by, say, 50%.

Addressing informational concerns involves the owner and the designer. The designer is often ill equipped to handle major increases in RFIs or changes without additional compensation. Whereas the contractor may be expected to work scheduled overtime, the same cannot be said for most design firms. The contractor needs to pressure the owner to apply pressure on the designer to keep pace with the contractor. Late submittal approvals and late answers to RFIs are detrimental. Lateness is a common theme in acceleration.

How the need for a change is identified is important. The worst-case scenario is the need for a change to be identified by the crew in the field. When a conflict is identified, the crew must stop work when it sees something that will not work. Idle time ensues as the crew waits for further instructions. It may be that the problem cannot be resolved on the spot, and the crew may be instructed to go work elsewhere. The supervisor has no opportunity to plan for work in another area. Planning is done "on the fly," resulting in more wasted time. At some time later, the change is issued. Now the crew returns to the work area at a later time, only to find that the environment has totally changed. One way the environment can change is that there now may be several other crews working in the same area; there is stacking of trades and congestion from other causes, leading to even more inefficient work hours.

How can this situation be avoided? Perhaps it is worthwhile to add someone to the project payroll to scour the plans and identify errors and omissions (Principle 6.1) in advance of the work being performed. This strategy will keep the crew in the field from finding the errors and omissions. These errors and omissions should be resolved before the crew goes to do the work.

Where numerous changes occur, it may be beneficial to organize a special crew to do changes to work that has already been installed (Principle 6.2). This method allows the production crew to continue production work. A word of caution is warranted. These two crews should not be too close to one another, as it can be demoralizing to the production crew to see their work being ripped out. A separate change crew also makes it easier to monitor change costs and claim these at the end of the job. On every change order, the contractor should include a disclaimer reserving the right to claim change costs at the conclusion of the job (Principle 6.3). Don't forget to comply with the notice clause in the contract (Principle 6.4). Failure to comply with these two principles may easily preclude the recovery of change costs.

A situation that has been observed on more than one occasion is that the designer abandons his or her responsibility for design. How does this happen? It usually occurs in relation to shop drawings that do not work, such as dimensional errors. The triggering response from the designer and owner may be to "make it work through the shop drawing process" or "the problem is one of means and methods." Either way, it then may fall to the contractor to do any redesign. So don't be surprised if the contractor has to perform some design work. The redesign may require a registered professional engineer.

All tools and equipment will experience greater use. This means that they will require more repair, so you need to increase the inventory of spare parts (Principle 7.1). If overstaffing is applied, you will probably need more tools. Sharing most tools is not a good idea (Principle 7.2). Sharing at a workstation, like a conduit bending station, is probably okay, but items like "come-along winches" are often hoarded, not shared.

Support Services Principles

Support services may be given little thought when acceleration occurs, but they are important and can inhibit good labor productivity. For the purposes of this chapter, support services are considered scaffolding, lifting capacity, and lifts, although support services can include many other items. Because of similarities with scaffolding, concrete formwork is also briefly discussed.

To erect more scaffolding, scaffolding must first be dismantled. Because of incomplete work, pending changes, and other factors, scaffolding may tend to remain in place longer. If this situation occurs, more scaffolding will be needed (Principle 8.1).

If the size of the workforce increases, more lifts may be needed (Principle 8.2). It may be more efficient if lifts are strategically assigned to an area rather than to a crew.

Because the goal is to produce more output in a shorter period of time, increased lifting capacity of all types may be justified (Principle 8.3).

Concrete formwork is similar to scaffolding in that before it can be used elsewhere, it must be stripped from some other location. Because curing is not accelerated, it may be necessary to put more formwork on the job (Principle 8.4). Certainly an evaluation is justified.

12.6 Case Study 1—Hyperion Wastewater Facility

The case study project is a wastewater project upgrade costing more than $1 billion. The project was a design-bid-build project. There were six major contracts.

This case study only covers one of the six contracts—the construction of three digester batteries.

Project Description

The contract was for the construction of three digester batteries consisting of 20 steel digester tanks. Fig. 12-2 shows a plan view of the site. Batteries D1 and D2 are in the center left of Fig. 12-2, and Battery E is in the upper right of the figure. Fig. 12-3 shows a cross section of one of the digester tanks. Fig. 12-4 shows batteries D1 and D2 under construction.

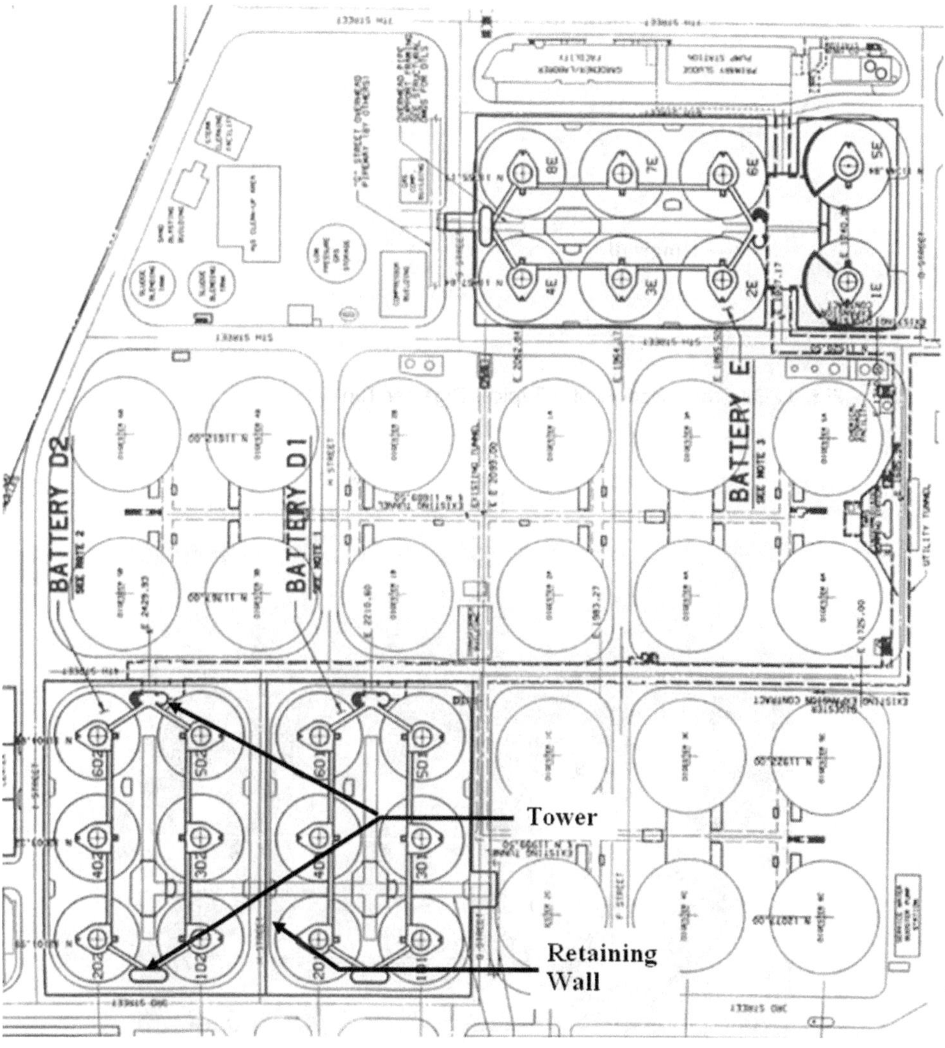

Fig. 12-2. Plan View of Hyperion Wastewater Digesters

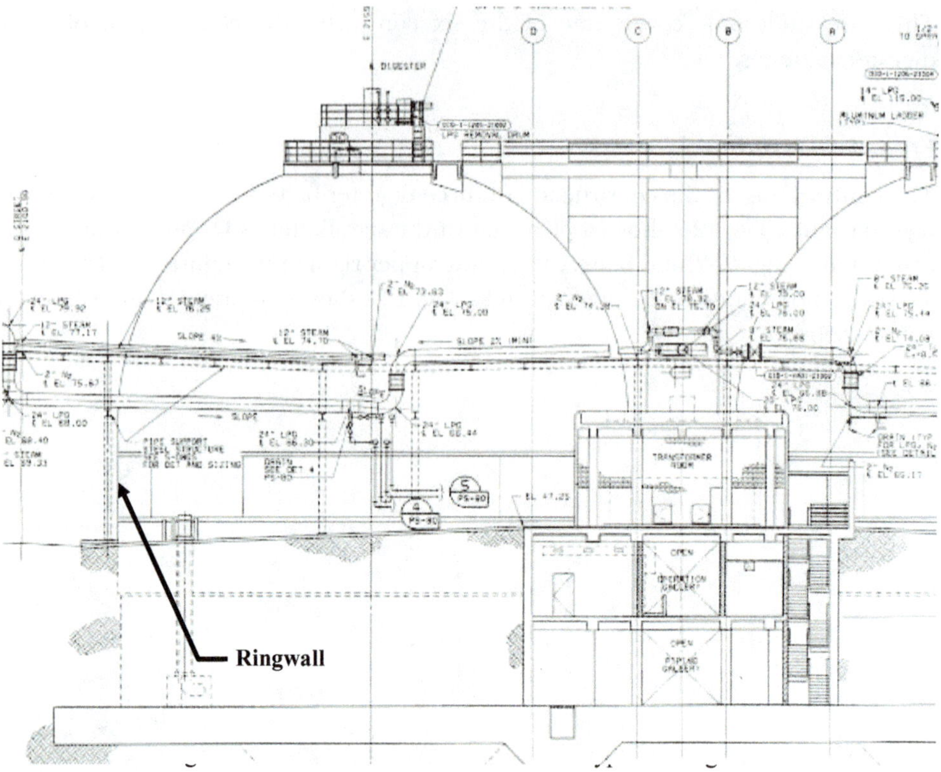

Fig. 12-3. Cross Section of a Typical Digester Tank

Fig. 12-4. Hyperion Site under Construction, Batteries D1 and D2

The original plan required by the contract called for the order of construction to be batteries D1–E–D2. The contractor requested and received permission to revise the order to batteries D1–D2–E. This change meant that the contractor avoided building the retaining wall between batteries D1 and D2, because both areas were actually excavated simultaneously (Fig. 12-5). The scope of work involved excavation; construction of foundations, cones, and mat slabs (Fig. 12-5); concrete ringwalls (Fig. 12-3); a below-grade piping gallery (Fig. 12-6); tank erection (Fig. 12-7); construction of pipe racks (Fig. 12-8) and related piping; and a gas dome on top of each tank (Fig. 12-9).

Fig. 12-5. Construction of Cones and Mat Slab for Batteries D1 and D2

Fig. 12-6. Below-Grade Piping Gallery

Fig. 12-7. Tank Erection

Fig. 12-8. Pipe Racks

Fig. 12-9. Gas Domes and Platform

Owner-Caused Delays

The wastewater authority was under a court order to cease its pollution, which meant that the project completion date could not be extended beyond December 1999 without penalty. But the project start date was delayed almost a year because of some unknown reason, thereby compressing the construction schedule. The authority further exacerbated the situation by not granting the contractor a time extension for many months. This placed the contractor in limbo and forced him to accelerate the schedule even though he was subsequently granted an extension of time.

Contractor-Caused Delays

The contractor was not without fault. He underestimated the difficulties he would have with the uniformly graded sand at the bottom of the excavation. He additionally underestimated the complexity of the towers at each end of the batteries (Fig. 12-2). The ringwall construction began inefficiently (Fig. 12-3).

Other Delays

Other delays affected the schedule. There were seals that leaked, requiring extensive rework. The defective seals were a design issue.

The contractor had difficulty with constructing the foundation cones that can be seen in Fig. 12-5. Groundwater infiltrated the first two cones of Battery D1. Because there was only one cone construction operation, the delay in D1 also delayed D2 and E.

The pipe hangers did not fit, and most had to be redesigned by the contractor. Contractor redesign was required because the authority told the contractor to correct dimensional errors through the shop drawing process. The hangers were somewhat complex, and valuable time was lost during the ensuing arguments, the redesign, and the lengthy approval process.

The Contract

The contract was written in such a way that many parts referenced numerous other parts. Thus, unrelated parts of the specifications were applied to other parts of the specifications. Some aspects were impossible to do. This situation led to many more arguments, which resulted in more wasted time. The authority made matters worse by deducting credits from the contractor's pay request when he could not comply. The contractor argued that the specifications were impossible, but to no avail.

Analysis

The Hyperion project demonstrates a number of common characteristics of schedule acceleration. These are the following:

- Delayed start,
- Not granting a time extension in a timely manner,
- Contractor- and owner-caused delays,
- Design errors,
- Contractor forced to assume design responsibilities,
- Late submittal approvals,
- Lack of schedule flexibility,
- Reduced payments or none at all, and
- Events beyond the owner's and contractor's control.

A labor inefficiency analysis was performed on the concrete formwork and piping. The estimated inefficient work hours by cause are shown in Tables 12-3 and 12-4.

As can be seen, there were numerous causes of the acceleration, and often in acceleration both owner and contractor share in the responsibilities for damages. From Tables 12-3 and 12-4, there were more than 135,000 inefficient work hours. Some were caused by the contractor and some by the owner. At a burdened rate of

Table 12-3. Inefficient Work Hours for Concrete Formwork by Cause

Issue	D1	D2	E	Total
Cone–cement	583	963	—	1,546
Undisturbed soil	1,051	582	707	2,340
Assemble forms	2,759	139	—	2,898
Schedule acceleration	12,689	4,813	7,007	24,509
Form disassembly	—	1,108	—	1,108
Well point system	—	—	7,084	7,084
Unexplained	10,578	4,146	4,607	19,331
Total	27,660	11,751	19,405	58,816

Table 12-4. Inefficient Work Hours for Piping and Supports by Cause

Issue	D1	D2	E	Total
Schedule acceleration	14,971	12,943	18,887	46,801
Gallery support redesign	1,122	780	—	1,902
Support finish issue	3,389	—	—	3,389
Late submittal approval	2,087	634	7,234	9,955
Temporary rack supports	—	3,988	—	3,988
Unexplained	1,941	3,378	4,982	10,301
Total	23,510	21,723	31,103	76,336

$38/h, the inefficient work hours amount to more than $5 million on just these two items of work alone.

12.7 Case Study 2—Determining the Amount of Formwork Needed

The following is a simplistic example for determining the amount of formwork needed and the effect of schedule acceleration. Suppose that a project plan is to employ a formwork crew consisting of eight carpenters. The plan is to work a 4-10 schedule. Assume that most of the formwork is wall formwork and that the budgeted performance of the crew is 0.065 wh/ft^2. Stripping will be done by a single four-person team of carpenters working all day on Tuesday.

Step 1—Figure Out How Many Teams Are Needed and the Sizes

There will be one formwork erection crew (eight carpenters) working 40 hours per week on a straight time schedule and 50 hours per week on an overtime schedule. There will be one stripping team, consisting of four carpenters. The plan is that one day each week (Tuesday), a four-carpenter team will erect formwork and another four-carpenter team will strip formwork. Therefore, 280 work hours are to be spent in erecting formwork and 40 work hours in stripping formwork.

Step 2—Figure the Weekly Quantity

Calculate the quantity of formwork erected per week for the crew when schedule acceleration is not a factor. To do so, calculate the total formwork erection and stripping work hours. Some 280 work hours will be spent in formwork erection and 40 work hours in stripping. Divide the erection and stripping work hours by the budgeted productivity, and divide this number by the appropriate rule of credit (see Table 7-2), 0.75 for erection and 0.10 for stripping.

$$\text{Weekly Formwork Erected (ft}^2) = \frac{280}{0.065 \times 0.75}$$
$$= 5{,}744\,\text{ft}^2$$

Using similar calculations, the four-carpenter team can strip 6,154 ft^2 in a day. Thus, the size of the surge pile needs to be 5,744 ft^2.

Step 3—Check Balance

Once a steady-state operation is achieved, the crew can erect formwork entirely from the surge pile of formwork that has been stripped. During any given week, a total of 5,744 ft^2 of formwork can be erected (based on 280 work hours) and a total of 6,154 ft^2 can be stripped (based on 40 work hours). Thus, the size of the surge pile needs to be 5,744 ft^2. The quantities should be closely balanced between erection and stripping; if not, adjustments should be made.

Step 4—Figure the Effect of Schedule Acceleration

Assume that the project has fallen behind schedule and the recovery plan is to work a 5-10 schedule using the same crew and crew assignments. How much more formwork should be procured? Summary data are given in Table 12-5.

Table 12-5. Summary Schedule

	4-10 Schedule @ 0.065 wh/ft² (ft²)	5-10 Schedule @ 0.065 wh/ft² (ft²)	5-10 Schedule @ 0.070 wh/ft² (ft²)	5-10 Schedule @ 0.085 wh/ft² (ft²)
Weekly formwork erected	5,744	7,385	6,857	5,647
Weekly formwork stripped	6,154	6,154	5,714	4,706
Size of surge pile	5,744	7,385	6,857	5,647

Many refinements can be made to the aforementioned methodology; however, the mathematics clearly show that there is more than a 25% increase in the quantity of formwork needed to accommodate the acceleration demands when going from 40 to 50 h/week. The productivity of the crew will degrade unless more formwork is procured. The size of the surge pile should be based on the high-value (erection) subtask. The size of the surge pile needed has increased from 5,744 ft^2 to 7,785 ft^2. But realize that the crew as originally organized cannot strip 7,385 ft^2, so more hours will have to be devoted to stripping. A new plan is required. The data also show the effect of productivity degradation. If the actual productivity worsens to 0.070 wh/ft^2, then the plan will have to be revised further. Note that if the productivity degrades to 0.085 wh/ft^2, the crew will erect less formwork than it was able to erect on a straight time (4-10) schedule.

This example shows that for overtime schedules of 50 work hours/week, the hours devoted to stripping are insufficient. The size of the surge pile will be too small, and productivity will degrade. A change in the crew assignment plan is necessary because the overall productivity of the formwork operation is sensitive to the efficiency of the operation.

References

Gilbane Building Company v. Two Turners' Electric Company. (2007). "Tex. App.–Houston [14 Dist.]." Providence, RI.

CHAPTER 13

Best Practices for Environmental Compliance

Compliance with environmental regulations is an important issue, and noncompliance can be costly for obvious and not so obvious reasons. This chapter gives general guidance because specific requirements vary from state to state. Best practices are cited in Table 13-1.

13.1 Why Compliance?

It is important to comply with environmental regulations for a number of reasons. The most obvious is that noncompliance can be costly. From a legal viewpoint, regulations have the same weight as statutes. Failure to comply with regulations is like failure to comply with a statute, and the fines for noncompliance can be significant (Fig. 3-12). More costs are incurred if a stop work order is issued by a regulatory agency. Most contracts require the contractor to comply with all laws, regulations, and local ordinances. Ignorance of laws and regulations is not a valid excuse. So noncompliance may also be a breach of contract.

A less obvious cost arises because some of the actions taken to ensure compliance also improve labor performance. Many of the actions discussed in Chapter 5 fall into this category. For example, a site drainage plan may lead to compliance with certain regulations and also improves labor performance (Figs. 5-4 and 5-5).

A third element of cost involves damage to the contractor's reputation. Noncompliance, fines, stop work orders, and court actions are often played out over many months in local newspapers. No contractor needs to be characterized as an unsatisfactory corporate citizen. A poor reputation for citizenship can lead to loss of business opportunities.

13.2 OSHA

The Occupational Safety and Health Administration (OSHA) is the main federal agency charged with enforcement of safety and health legislation. The agency has

Table 13-1. Best Practices for Environmental Compliance

No.	Practice
	General
1.1	Be sure a wetland permit has been obtained (if required) before bidding.
1.2	Become knowledgeable about all federal, state, and local laws, ordinances, and regulations.
	Environmental Compliance Plan
2.1	Integrate the environmental compliance plan with the site drainage and erosion and sediment control plans.
2.2	Realize that as the project evolves, the requirements may change.
2.3	Include maintenance of facilities and schedule in the plan.
	Structural Work
3.1	As the work progresses, the nature of the work evolves from drainage and excavation issues to trash removal issues.

published numerous regulations for the construction industry. Despite often being maligned, the agency has overseen substantial improvements in the safety record of the construction industry since its creation around 1970. The OSHA regulations go beyond detailing what a contractor must do to create a safe work environment; there are also reporting requirements. Contractors must report to OSHA all fatalities and lost-time accidents.

13.3 The Wetland Permit

A wetland permit is required any time one is operating in a wetland. Wetland permits are issued to owners by the U.S. Army Corps of Engineers. In general, the permit provides that for every acre of wetland disturbed, an acre of new wetland must be created. A wetland is not necessarily a marshy area. A wetland is defined by the wildlife habitat, soil type, and other criteria. The regulatory requirements may vary according to the type of facility being constructed.

13.4 Other Requirements and Regulations

It is impractical to discuss various regulatory requirements because the requirements are unique to each state. In general, all states publish requirements for erosion and sediment control, discharges, and waste control (hazardous and otherwise). Most requirements are mandatory, and the damage to a contractor's reputation can be significant (Fig. 3-12). A good place to begin is the Construction Industry Compliance Assistance Center at http:///www.cicacenter.org/cs.cfm. Upon entering this website, one is asked (1) the state, (2) type of construction, and (3) potential impact.

For the input, VA requires land development and wildlife habitat and stream or other water body. The following output illustrates the nature of useful information provided.

Stormwater and Erosion Control

As of March 10, 2003, in order to discharge stormwater from a construction site, all construction projects that disturb 1 acre or more of land must have either:

- an individual stormwater permit, or
- coverage under one of Virginia's general permits.

Disturbance includes, but is not limited to soil disturbance, clearing, grading, and excavation. Operators of sites disturbing less than one acre are also required to obtain a permit if their activity is part of a "larger common plan of development or sale" with a planned disturbance of one acre or greater.

To apply for an individual permit or for coverage under a statewide general permit you must file an application with the Department of Environmental Quality prior to commencement of your construction activities.

In addition to these statewide rules, you may be required to meet additional local stormwater and erosion control regulations. Check with your city or county government to determine if additional local rules apply to your construction project.

Permit Application Forms and Instructions for Virginia:

- Notice of Termination
- List of 2014 Construction GP Complete Registration Statements

General Permits:

- VAR10 General Permit for Discharges from Construction Activities
- Stormwater Program Contact Information

Additional Stormwater Compliance Resources:

- Engineers' Toolkit: Virginia Stormwater Management Program (VSMP) Permit Regulations

Species Incidental Take Permits

Private landowners, corporations, state or local governments, or other nonfederal landowners who wish to conduct activities on their land that might incidentally harm (or "take") a species listed as endangered or threatened must first obtain an incidental take permit from the U.S. Fish and Wildlife Service.

To obtain a permit, the applicant must develop a Habitat Conservation Plan (HCP) designed to offset any harmful effects the proposed activity might have on the species. The HCP process allows development to proceed while promoting listed species conservation. The "No Surprises Policy" provides assurances to landowners participating in HCP efforts.

For assistance or to inquire about permits, contact your U.S. Fish and Wildlife Service Office(s):

- Northeast Region Fish and Wildlife
- Virginia Ecological Services Field Office

Additional state regulations may apply. For information on state rules relating to endangered species contact the Virginia's Department of Game and Inland Fisheries. Also, look below under "Additional Information" for links to applicable regulations.

Additional Information for Virginia:

- Virginia Wildlife Action Plan

Hazardous Waste

Wastes are classified as "hazardous" if they have certain dangerous properties (for instance, if they are flammable, corrosive, or toxic). Examples of hazardous wastes commonly generated at construction sites include: asbestos, lead (e.g., in paint), solvent/thinner, fuel, stains, unused paint, oil/lubricants, compressed gas cylinders, and unpunctured aerosol cans.

EPA defines three categories of hazardous waste generators based upon the quantity of hazardous waste they generate per month. Knowing your waste category is important because it can affect which regulations apply to your operation. The categories are:

- Conditionally exempt small quantity generators (CESQGs), which generate less than 220 lb (100 kg) per month.
- Small quantity generators (SQGs), which generate between 220 lb (100 kg) and 2,200 lb (1,000 kg) per month.
- Large quantity generators (LQGs), which generate more than 2,200 lb (1,000 kg) per month.

Here are the main rules:

- Any solid waste (which may be a solid, sludge, liquid, or contained gas and is any spent, used, discarded, abandoned, or no longer usable material) generated by your business must be evaluated to determine if it is a hazardous waste.
- You must accumulate hazardous waste in containers (e.g., DOT approved drums) or tanks. If you are a SQG or LQG, the containers must be labeled with the words "Hazardous Waste" and the start date of accumulation must be shown.

- SQGs and LQGs may store the waste on site for up to 180 days (270 days if the waste is shipped to a facility more than 200 miles away). There is no time limit on how long CESQGs may accumulate hazardous waste. However, they may never exceed the 1,000-kg maximum quantity limit.
- You have to use a licensed waste hauler to transport the waste off your site, and it must be sent to a facility permitted to accept that type of waste. (CESQGs may transport the waste in their own vehicles if certain rules are followed.)
- You have to keep a record (called a "shipping manifest") of the shipment for a specified period of time. (CESQGs are not required to use the Uniform Hazardous Waste Manifest when shipping waste off site. However, they must maintain records demonstrating that they have properly disposed of their wastes.)

A special note on "empty containers." Once they meet the definition of empty, containers with small amounts of hazardous residues are generally not considered to be regulated hazardous waste. They are regulated as a solid waste when disposed. A container is considered "empty" when all wastes have been removed by using common industrial transfer practices and no more than one inch of residue remains in the container (or <3% of the volume). If the container is >110 gallons, then it may contain no more than 0.3% of residue. Containers holding compressed gases are considered empty when the gas product contents have been used and the container is at atmospheric pressure. However, aerosol cans using a compressed gas propellant are considered empty when the container meets the empty condition for the product contents. For example, a spray can that is full of solvent-based paint or ether starting fluid, but that has no remaining propellant gas pressure, may still be considered a hazardous waste. Releasing a compressed hazardous gas to the atmosphere or allowing container contents to evaporate to make the container empty is NOT allowed.

- Virginia hazardous waste permits and supporting resources
- Virginia contact information for hazardous waste.

Spill Prevention and Countermeasure Plans

The rule includes spill prevention and countermeasure plans for spills from aboveground and certain underground storage tanks. A construction site project must meet SPCC regulatory requirements if it meets the following three criteria:

- It stores, uses, transfers, or otherwise handles oil;
- It has a maximum aboveground storage capacity greater than 1,320 gallons of oil (which includes both bulk and operational storage volumes) OR total underground storage capacity greater than 42,000 gallons of oil; AND

- There is a reasonable expectation (based on the location of your site) that an oil spill would reach navigable waters or adjoining shorelines of the U.S.

For more information, go to CICA SPCC Guidance.

Air Pollution Control

Certain activities at construction sites can affect air quality, including operation of diesel engines, open burning, land clearing, and demolition. Air pollution rules applicable to construction sites are contained in federal, state, and local regulations and ordinances:

- Regulations restricting or prohibiting open burning are typically written and enforced by state, regional, and/or local governments. These rules are aimed at reducing the generation of smoke, preventing smoke and ash from affecting people and the environment, and preventing wildfires.
- Airborne dust generated from construction activities may be regulated by regional and local rules. Typical rules require builders to implement best management practices (e.g., establish vegetation, watering, sequential clearing, wheel washing, reducing vehicle speed, prompt removal of dirt/mud tracked onto highways) as abatement measures.
- On a national basis, the strategy for controlling air pollution from diesel engines involves low-pollution requirements for new diesel engines and rules covering the fuel used by these engines. EPA has issued emission standards for new, nonroad diesel engines, such as construction and farm equipment, and is working to strengthen these standards in the future. Some air districts regulate emissions from portable equipment that have spark ignition or diesel-fired piston type internal combustion engines that may be used at construction sites, including generators, pile drivers, compressors, welders, cranes, well drilling, etc.

To determine which air pollution regulations apply to your construction project, contact your local air district or state air pollution control agency:

State Air Agency

- Virginia Department of Environmental Quality, Air Quality Division Air Regulations

Local Agencies

- Alexandria Health Department, Office of Environmental Quality
- Air Permits
- Arlington County Department of Environmental Services
- Fairfax County Health Department, Division of Environmental Health
- Roanoke Environmental Health Division

Asbestos

Asbestos is used in certain products where heat and corrosion resistance is important, such as building materials and vehicle brakes. For building materials, asbestos is most often found in products such as shingles, floor tiles, cement pipe and sheet, roofing felts, insulation, ceiling tiles, fire-resistant drywall, and acoustical products. You are particularly likely to find asbestos in older materials in existing buildings.

Asbestos is known to be a health hazard. If inhaled, it accumulates in the lungs, creating scar tissue that makes it more difficult to breathe, and leading to a type of lung cancer called mesothelioma, which is almost always fatal by the time it is diagnosed. Therefore, asbestos is extensively regulated. For the construction industry, there are two primary concerns, each having its own set of rules:

- EPA (U.S. Environmental Protection Agency) established rules under the authority of the Clean Air Act (CAA) to protect the general public from the release of airborne asbestos fibers during the demolition or renovation of buildings. These rules are referred to as the Asbestos NESHAP (National Emissions Standards for Hazardous Air Pollutants). They cover work practices to be followed during demolition and renovation and other activities involving the processing, handling, and disposal of asbestos-containing material.
- OSHA (Occupational Safety and Health Agency) and EPA established rules to protect the health of workers who are most likely to contact asbestos. These rules established strict worker exposure limits and set out requirements for employers regarding exposure assessment, medical surveillance, recordkeeping, and hazard communication. The EPA rules cover state and local government employees, and the OSHA rules cover private sector workers.

Virginia Asbestos Web Page: None located.

Virginia Asbestos Contact:
Joe Kossan
Va Dept of Professional and Occupational Regulation
Voice: 804-367-8595
Fax: 804-367-2475
Asbestos@dpor.state.va.us

EPA Contact:
Tia Chambers
U.S. EPA, Region 3
1650 Arch Street
Mailcode: 3AWC32
Philadelphia, PA 19103
Telephone: (215) 814-2164
Telefax: (215) 814-3113

OSHA Program Information

OSH Consultation Program
Virginia Department of Labor & Industry
Occupational Safety & Health
Training & Consultation
13 South 13th Street
Richmond, Virginia 23219
(804) 786-6613
(804) 786-8418 FAX
E-mail: wer@doli.state.va.us

Construction and Demolition Debris

Construction and demolition (C&D) debris refers to materials produced in the process of construction, renovation, and/or demolition of structures. In most cases, C&D debris is nonhazardous and is regulated by states and local governments rather than by EPA. An exception would be where C&D debris contains hazardous waste, such as removed asbestos insulation.

C&D debris is a significant issue in the U.S. because of the enormous volume of C&D debris generated. A large fraction of C&D debris ends up in municipal solid waste landfills or in special C&D landfills, which may have the potential to contaminate groundwater. Also, each year, there is less land available for waste disposal. As a result, many state and local governments are seeking ways to divert C&D debris from land disposal, including the promotion of recycling. Also, Green Building programs exist where the focus is on minimizing the generation of wastes.

State and local regulations may limit where you can dispose of C&D debris. For example, some local governments do not permit C&D debris to be disposed of in their municipal landfills. Also, some local governments, particularly in California, require construction companies to recycle a minimum percentage of the C&D debris generated.

Use the links below to locate regulatory information and other compliance assistance and P2 resources for Virginia. It is also recommended that you contact your city or county government to determine if local rules also apply to your project.

Primary State Agency: Department of Environmental Quality, State Solid Waste Page

- Construction and Demolition Debris Main Page
- Construction and Demolition Debris Regulations
- Landfill List

Universal Wastes

Universal wastes are hazardous wastes that are more common and pose a lower risk to people and the environment than other hazardous wastes.

Federal and state regulations identify universal wastes and provide simple rules for handling, recycling, and disposing of them.

In Virginia the Department of Environmental Quality enforces the hazardous waste regulations, including the universal waste rule. If you have any questions, contact someone at Solid and Hazardous Waste Regulatory Programs.

Virginia's Universal Waste Rule covers the following wastes: batteries (e.g., certain lead-acid batteries not recycled under other regulations; button silver-oxide and zinc-air; and 9-volt, C, AA, coin, and button rechargeable lithium); pesticides; mercury-containing devices (e.g., thermostats, switches); and electric lamps (e.g., fluorescent, high-intensity discharge, sodium vapor, and mercury vapor). Certain other battery types (e.g., alkaline and carbon zinc cells that have been manufactured without mercury) may not be classified as hazardous wastes and would therefore not fall under the Universal Waste Rule. However, as a matter of responsible practice, such batteries should be collected and sent to recycling facilities, rather than being landfilled.

To learn more about how universal wastes are managed, visit CICA's Universal Waste webpage for Virginia.

Lead

On March 31, 2008, to better protect children from lead-based paint hazards, EPA issued a new rule for home improvement contractors and maintenance professionals who renovate or repair pre-1978 housing, child care facilities, or schools. The rule requires that by April 2010 contractors and maintenance professionals be certified, that their employees be trained, and that they follow protective lead-safe work practice standards. More information.

Lead-based paint has been banned since 1978, but many older structures still have this paint on walls, woodwork, siding, windows, and doors. Construction and demolition workers can be exposed to lead contamination by cutting, scraping, sanding, heating, burning, or blasting lead-based paint from building components, metal bridges, and metal storage tanks. In addition to exposure to workers, lead-based paint debris or dust can also make its way into soil, potentially contaminating surface waters. Lead poisoning is a serious health threat for adults and is especially damaging to young children.

To learn more about lead regulations related to demolition and construction, visit CICA's lead paint abatement webpage for Virginia.

Mercury

Construction, remodeling, and demolition contractors often come into contact with used equipment or products that contain mercury. Some common examples include thermostats, switches and relays, fluorescent lamps, and batteries. When mercury-containing equipment or products are removed from service, they may meet the definition of a hazardous waste, and in such cases, state and/or federal hazardous waste rules apply to the management and disposal of those items. In some cases, these wastes may be managed as "universal wastes," which have less complicated rules for handling and recycling or disposal.

To learn more about mercury regulations related to demolition and construction, visit CICA's mercury webpage for Virginia.

OSHA

The federal Occupational Health and Safety Administration (OSHA) operates the primary job safety and health program in twenty-nine (29) of the fifty states. This includes conducting inspections and enforcing its standards. States with approved programs must set job safety and health standards that are "at least as effective as" comparable federal standards. In most cases, states adopt standards identical to federal ones.

Consultation services are available in every state. In most cases, these are free and are conducted at your work place or building site. These services help employers identify and correct workplace hazards and can help you avoid violations and penalties.

To learn more about OSHA rules, visit CICA's OSHA webpage for Virginia.

The National Environmental Policy Act of 1969 (NEPA)

NEPA marked the beginning of the environmental review process for all federal actions, including the construction of highway and bridge projects falling under the Federal Highway Administration (FHWA). NEPA established a mandate for federal agencies to consider the potential environmental consequences of their proposed actions, document the analysis, and make this information available to the public for comment prior to implementation.

In September 2002, President George W. Bush signed Executive Order 13274. Titled "Environmental Stewardship and Transportation Infrastructure Project Reviews," the order directed federal agencies like the U.S. Army Corps of Engineers and EPA to collaborate more effectively to streamline the environmental review and development of transportation infrastructure projects and advance major transportation projects. Read more about NEPA requirements.

General Virginia Information

In this section, you will find links to state agencies and regulations.

- Virginia's Home Page
- Primary Environmental Enforcement Agency: Department of Environmental Quality

- Regulations Online: General |Air |Water |Solid/Hazardous Waste
- Air Agency: Air Quality
- Water Agency: Water Quality
- Hazardous Waste Agency: Solid and Hazardous Waste Regulatory Programs
- Contacts: Ombudsman and other SBAP contacts

Note that the CICA Center website contains much general information about a variety of other permits and regulatory requirements besides the wetland permit.

13.5 Best Practices

General

As we see in Table 13-1, be sure that a wetland permit has been obtained by the owner (if required) before bidding (Practice 1.1). This permit is issued by the U.S. Army Corps of Engineers and should be obtained by the owner. Procuring this permit can be a lengthy process. If the permit has not been obtained, the contractor should seriously consider not bidding on the project. Instances have been observed where the contractor has been directed to proceed without the owner procuring the required right-of-way (highway and railroad) permits. The unscrupulous owner then sued the contractor for lack of performance. Regardless of the judicial outcome, the contractor will lose money.

Become knowledgeable about all federal, state, and local laws, ordinances, and regulations (Practice 1.2). Ignorance of the regulation is no excuse. Be sure that the responsible person in charge at the site is also knowledgeable.

Environmental Compliance Plan

Integrate an environmental compliance plan with the site drainage and erosion and sediment control plans (Practice 2.1). Consider integrating the three plans into one.

Realize that as the project evolves, the requirements may change (Practice 2.2). Plans made for the site preparation and excavation phases may be of limited use for the later phases of the project.

Include maintenance of facilities and the schedule in the plan (Practice 2.3). A maintenance plan needs to indicate how often facilities need to be repaired or rehabilitated, and a schedule needs to indicate when new facilities or actions are to be applied.

Structural Work

As the work progresses, the nature of the work evolves from drainage and excavation issues to trash removal issues (Practice 3.1). Trash removal is an important part of ensuring good labor performance. But environmental considerations may dictate that trash may need to be segregated. Landfills may not accept all forms of waste. Separate disposal practices may be needed for paints, oil, grease, and other products.

CHAPTER 14

Productivity Evaluation

Throughout this text, principles have been presented that when applied improve labor productivity and lower cost. It follows that productivity can be evaluated based on how well fundamental principles are applied. This chapter presents a procedure for doing so.

Many times, problems at the site that negatively affect labor productivity are readily obvious but persist nevertheless. What the project manager and project engineer need is a quick way to evaluate the site to determine if improvements to productivity can be readily made. The following is a quick productivity evaluator.

Consider an example. Suppose a crew of 10 pipefitters is installing process pipe on an industrial site. The crew is in need of 6-in. stainless steel pipe spools, so the supervisor sends four of the crew members to retrieve spools so that the crew can continue installation. Unfortunately, the spools cannot be located. Soon thereafter, the remaining six pipefitters run out of work and are idle, waiting for more materials. The material was not preloaded (Table 6-1, Principle 3.3) or marked for easy identification (Table 6-1, Principle 1.3). In this scenario, not following some of the fundamental principles in Table 6-1 is the root cause of productivity loss; the idle time is merely the symptom.

14.1 Quick Evaluator

A quick project evaluator can be developed using various aspects of the work that are readily observable. First, conduct a tour of the site and answer the following questions. The common aspect of these five questions is that each condition is readily observable.

Yes	No		
☐	☐	1.	Is the housekeeping unsatisfactory (focus on trash, clutter, etc.)?
☐	☐	2.	Do you observe much idle time?
☐	☐	3.	Do you observe instances of overstaffing (for example, five workers installing bulkheads)?
☐	☐	4.	Do you observe workers carrying materials up the stairs, one component at a time?
☐	☐	5.	Is there significant congestion observed? (Try to ascertain the cause; it may be stacking of trades caused by out-of-sequence work or material storage practices.)

If the answer to any of these questions is yes, then significant improvements in labor productivity are possible and a detailed analysis is likely justified.

14.2 Detailed Analysis

The preceding chapters have outlined multiple principles to apply during the planning and execution of a construction project. It follows that these principles can be used to evaluate the management practices on a project and to highlight opportunities for improvement. The principles are summarized in Table 14-1. Duplicate principles have been deleted, and schedule acceleration and subcontractor management principles have not been included.

It is probably easiest if only one activity is evaluated; however, the entire site may need to be evaluated. If one activity is observed, concentrate on critical activities or an intense activity.

Each principle is scored on a scale of 1 to 10. But the scoring is often imprecise. Therefore, using the principles cited in the preceding chapters, the following scheme is applied:

Score	Description
0	The principle is never observed.
5	The principle is infrequently applied or inconsistently observed.
10	The principle is always applied or consistently observed.

Each applicable principle is scored. Not all principles are applicable to every project. For example, the principle "Leave snow in place as insulation until ready to excavate" would not be applicable to Florida projects or to a building project. Therefore, not all principles will be evaluated, nor will all principles be observed. For example, you may not observe when or how often cranes are relocated.

Sum the total of the scores from the principles evaluated and divide this total by the number of principles evaluated. This yields an average score per principle. Experience with this technique has shown that typical scores are in the range of 2 to 8. The better managed jobs have higher scores. Principles with individual scores of 0 or 5 are candidates for that principle to be applied.

14.3 Significant Fundamental Principles

Practicing engineers may find that 90-some fundamental principles are too cumbersome. In a recent study of drywall construction practices (Thomas 2013), the following significant fundamental principles were identified:

- Maintain good housekeeping.
- Distribute materials throughout the floors and rooms of the facility (preload).

Table 14-1. Project Evaluations

No.	Principle	Score 0	5	10
	Planning			
1.	Make effective use of constructed and footprint areas.	☐	☐	☐
2.	To allow more room and to mitigate rain runoff, install utilities and permanent site drainage first.	☐	☐	☐
3.	Minimize storage area needs by off-loading directly from the delivery truck.	☐	☐	☐
4.	Minimize storage area needs, enhance the schedule, and improve labor performance by preloading materials onto each floor or area.	☐	☐	☐
5.	Make use of service elevators to transport lightweight or bulky items like duct and drywall to each floor or area.	☐	☐	☐
6.	Develop multiple site layout plans.	☐	☐	☐
7.	Sequence activities so as to avoid making noncritical activities critical.	☐	☐	☐
8.	Execute activities concurrently, rather than sequentially.	☐	☐	☐
9.	Minimize the number of times a crane has to be moved.	☐	☐	☐
10.	Unless necessary, do not plan the same activity to occur on the same day of subsequent weeks, for example, we plan to place concrete every Thursday.	☐	☐	☐
	Weather			
11.	Use a 4-10 work schedule to permit a makeup day on straight time.	☐	☐	☐
12.	Use annual cycles to schedule around trades most affected by weather.	☐	☐	☐
13.	Where justified, accelerate the schedule to avoid winter cost.	☐	☐	☐
14.	Where possible, reserve some work that can be done on inclement workdays.	☐	☐	☐
15.	Enclose and seal buildings for weather protection.	☐	☐	☐
16.	Seal exposed areas each day.	☐	☐	☐
17.	Use drainage systems and sump pumps and plan for runoff, especially from off site.	☐	☐	☐
18.	Make use of permanent site drainage by installing drainage facilities early.	☐	☐	☐
19.	Apply an adequately engineered working surface.	☐	☐	☐
20.	Plow wet ground to accelerate drying.	☐	☐	☐
21.	Build and maintain all-weather roads.	☐	☐	☐
22.	Leave snow in place as insulation until ready to excavate.	☐	☐	☐
23.	Shift work hours to avoid the heat of the day.	☐	☐	☐
24.	Provide break trailers for relief from cold or heat.	☐	☐	☐
25.	Protect materials from weather.	☐	☐	☐

(continues)

Table 14-1. Project Evaluations *(Continued)*

No.	Principle	Score 0	5	10
26.	Store materials on timbers and pallets to keep them out of mud.	☐	☐	☐
	Materials			
27.	Do not store materials close to the building.	☐	☐	☐
28.	Locate parking areas, toolsheds, trailers, and spoil piles as far from the building as practical.	☐	☐	☐
29.	Mark stored materials so that they can be readily distinguished from similar materials.	☐	☐	☐
30.	Materials should be stored to permit easy access and retrieval.	☐	☐	☐
31.	Make effective use of surge piles (stockpiles) to ensure that work (components) are always available for the crew.	☐	☐	☐
32.	Avoid multiple staging areas at the site because this can lead to double-handling of materials and inefficiencies when unloading materials at the site.	☐	☐	☐
33.	Whenever possible, especially if storage space is limited, consider final staging of large materials at an off-site location.	☐	☐	☐
34.	Reserve areas next to the building for material deliveries or materials being moved to the work face areas.	☐	☐	☐
35.	Backfill around the building as soon as practical to permit the area to be used as a staging area.	☐	☐	☐
36.	The amount of material stored inside should be kept to a minimum.	☐	☐	☐
37.	Preassemble components into larger components or subassemblies.	☐	☐	☐
38.	Integrate the sequence of work with the storage plan so that interior space can be used without interfering with the work.	☐	☐	☐
39.	Ancillary tasks like unpacking, cutting, reshaping, and preassembly should be done away from the work face when practical.	☐	☐	☐
40.	Maintain good housekeeping.	☐	☐	☐
41.	Ensure that deliveries are properly sequenced to be consistent with the work plan.	☐	☐	☐
42.	Make sure that the delivery rate from vendors is compatible with the installation rate of the field crew.	☐	☐	☐
43.	Avoid the use of earthen ramps into below-grade areas.	☐	☐	☐
	Workforce Management			
44.	Staff the activity with labor resources that are consistent with the amount of work available to be performed. This includes taking adequate account of the variability in the project.	☐	☐	☐

Table 14-1. Project Evaluations *(Continued)*

No.	Principle	Score 0	Score 5	Score 10
45.	Have a good termination or layoff policy at the crew and project levels.	☐	☐	☐
46.	In instances of uncontrollable variability in workload, more labor than planned may need to be applied rapidly to complete work in the required time frame.	☐	☐	☐
47.	The concept of a working supervisor is not a good idea.	☐	☐	☐
48.	Share with the crew expectations in terms of production, work hours, time, and completion dates.	☐	☐	☐
49.	Each crew must be allowed to work at an unimpeded pace to maximize labor efficiency.	☐	☐	☐
50.	Daily schedules should be planned to prolong the periods of the most productive work activity.	☐	☐	☐
51.	Make the primary focus of the crew's work directed to high-value work. Never stop working on high-value work.	☐	☐	☐
52.	Efficient material handling and timely deliveries are important for good productivity, especially on labor-intensive operations.	☐	☐	☐
53.	Work on low-value subtasks concurrently with high-value work.	☐	☐	☐
54.	Perform incidental and cleanup work concurrently with high-value work.	☐	☐	☐
55.	A crew should be viewed as a collection of flexible size work teams.	☐	☐	☐
56.	Minimizing or eliminating disruptions improves performance.	☐	☐	☐
57.	Keep the work face area free of trash and clutter.	☐	☐	☐
58.	Create multiskilled crews with individuals who have different skills, who, when combined, form the multiskilled crew.	☐	☐	☐
59.	Where concurrent multiskilling tasks are applied, crew assignments need to consider carefully the size of the teams, the production output to be expected for each team, and how many hours the teams should be given to accomplish their work.	☐	☐	☐
60.	Preassemblies and modules are an effective way to reduce the field labor component of the installation activity.	☐	☐	☐
61.	Where possible, avoid symbiotic activities.	☐	☐	☐
62.	There should be adequate time buffers between each activity.	☐	☐	☐
	Activity Sequencing			
63.	Make schedule allowances for approval of submittals, fabrication, and delivery, especially for long-lead items.	☐	☐	☐
64.	Develop a submittal schedule showing due dates and enforce the schedule.	☐	☐	☐

(continues)

Table 14-1. Project Evaluations *(Continued)*

No.	Principle	Score 0	Score 5	Score 10
65.	Consider economies of scale for pertinent components and propose pertinent changes to the designer.	☐	☐	☐
66.	Sequence activities to eliminate work tasks.	☐	☐	☐
67.	Sequence activities to avoid congested work areas.	☐	☐	☐
68.	Sequence activities to minimize the number of critical path activities.	☐	☐	☐
69.	Complete critical path activities before starting noncritical activities.	☐	☐	☐
70.	Where possible, do activities concurrently.	☐	☐	☐
71.	On slow activities, use multiple workstations.	☐	☐	☐
72.	Backfill around the facility as quickly as possible.	☐	☐	☐
73.	Use permanent windows in lieu of a temporary (polyethylene) enclosure.	☐	☐	☐
	Trade Sequencing			
74.	The work in the corridor ceiling (bulk installation) should be done first.	☐	☐	☐
75.	For maximum labor efficiency, the partition walls in a particular area should not begin until the ceiling work in the area is complete.	☐	☐	☐
76.	For maximum labor efficiency, the ceiling grid in a particular area should not be started until the ceiling work in that area is complete.	☐	☐	☐
77.	Develop a sequence plan detailing the order in which the floors or broad areas will be worked and how the work on a single floor or area will be completed.	☐	☐	☐
78.	Engage all affected subcontractors in developing the sequence plan.	☐	☐	☐
79.	Begin work in the corridor and then progress to individual rooms.	☐	☐	☐
80.	Install bulky, heavy items first and smaller field-run items last.	☐	☐	☐
81.	Allow ample time lags between activities.	☐	☐	☐
82.	Ensure that there is ample backlog of work so that crews and teams can establish their own pace of work.	☐	☐	☐
83.	Do not start work in an area unless all the work can be done in one pass through the area.	☐	☐	☐
84.	Size crews and teams so that different trades work at roughly the same pace.	☐	☐	☐
85.	Aggressively manage subcontractors.	☐	☐	☐
86.	Plan efficient methods for moving materials to their installation locations.	☐	☐	☐

Table 14-1. Project Evaluations *(Continued)*

No.	Principle	Score 0	Score 5	Score 10
	Congestion			
87.	Select vendors near the site (minimum travel distance) to better control delivery schedules.	☐	☐	☐
88.	Where possible, outsource work to others and use preassemblies to move as much work off site as practical.	☐	☐	☐
89.	Consider putting the project office in a part of the facility being constructed, and avoid moving the office once it has been located.	☐	☐	☐
90.	Move the project office into place only after the critical path work in that location is finished.	☐	☐	☐
91.	Where possible, schedule all noncritical path work at times so as not to interfere with critical path work.	☐	☐	☐

- Have a good work plan.
- Don't hang board until the permanent windows have been installed.
- Sequence the work to minimize "callback" time.
- Do not insist on working supervisors.

14.4 Selecting Significant Fundamental Principles

Significant fundamental principles were identified as follows. Two projects were selected as representing superior management, and two projects were selected to represent subpar management. The superior project subset was studied in detail, and the practices that were consistently applied (score of 10) were identified as the good practices subset. Next, the subpar projects were studied to identify which practices from the good practices subset were not applied on either subpar projects (score of 0). A practice that was consistently followed on both best managed projects (score of 10) and not followed on either subpar managed projects (score of 0) was identified as a "significant fundamental principle." Because each circumstance is unique, the procedure may be revised to suit the particular situation.

References

Thomas, H. R. (2013). "Benchmarking interior system contractors (ISCs)." *Pract. Period. Struct. Des. Constr.*, **18**(3), 04014048.

Glossary

Bid shopping—The practice of a prime contractor after being identified as the lowest bidder going back to subcontractors and asking them to reduce their quote.

Buffers—An item or action that allows separation of several activities. A buffer allows a crew to proceed at an unimpeded pace.

Bulk installation—The middle period of time during an activity when contractors are installing the easiest work, for example, long straight runs of pipe.

Controlling subcontractor—A subcontractor who controls the pace of the work of other subcontractors because their installation rate is slower or they are first in a given area. If the controlling subcontractor falls behind schedule, other subcontractors are delayed too.

Critical path—A string of activities in a critical path method (CPM) schedule where there is no schedule flexibility or latitude. Any delay to one of these activities results in a delay to the completion of the project. All activities on the critical path are called critical activities.

Driving activity—Sometimes called a controlling activity, this activity establishes the pace of following activities or controls the attainment of a milestone date.

E&S plan—An erosion and sediment (E&S) plan defines how runoff is to be controlled so as not to harm local streams. In many locales, an E&S plan is a regulatory requirement.

High-value work—Work that requires significant labor content per unit. High-value work may be a pay item in the schedule of values. For example, formwork erection, steel erection, and duct erection are high-revenue-producing subtasks. Hanger installation, tightening of bolts, and formwork stripping are not high-revenue subtasks.

Incidental work—Work that is necessary but for which there is usually no payment. There are usually few work hours involved. Housekeeping is an example of incidental work.

Laydown area—An area where materials may be stored for an extended period of time. These areas are usually associated with material deliveries when deliveries are made in large lot sizes. For example, if all the structural steel is delivered at a certain time in the project schedule, it is stored in a laydown area.

Leisure time schedule—A schedule, often a bar chart, that is based on conventional means, methods, and equipment.

Low-value work—Work that requires small amounts of labor content per unit. For example, plumbing and stripping of formwork, tightening of steel bolts, and installing nonseismic duct hangers are low-value-producing subtasks. Low-revenue work does not imply that the work is not necessary.

Measured mile—A method of quantifying labor inefficiencies where the labor productivity during a period of time in which the contractor's performance is affected by owner-caused events is compared with the labor productivity during a period of time when performance was affected by events caused only by the contractor him- or herself. This latter period is called the measured mile.

Multiskilling—The work of a construction crew that possesses a broad range of skills, giving the crew the capability of performing a wide range of tasks, such as formwork erection, steel reinforcement installation, and concrete placement.

Preloading—The practice of locating materials at the place where they are to be installed or used. Preloading involves locating only the right amount of materials, no more and no less, at the location where they are to be used. Preloading means that the crew does not have to travel any distance to retrieve materials.

Productivity—The work hours required to produce a finite amount of work divided by the crew work hours needed to produce this work. With this definition, higher numbers show poorer performance.

Rules of credit—Earning rules for subtasks used to give credit for work partially complete. They are unitless numbers based on the relative amount of labor effort required to complete a single unit of work.

Schedule acceleration—When contractors are required to complete more work in a unit of time than was planned when the bid was submitted.

Shakeout—The practice of sorting through delivered materials to make sure that the correct number of correct components were delivered and that the components are the correct size, have been fabricated correctly, and have not been damaged.

Short interval production schedule (SIPS)—A type of schedule covering a relatively short period. This schedule is not a look-ahead schedule because there is no element of hope in the schedule, that is, "I plan (hope) to do this in the next two weeks." The SIPS represents a commitment that the work prescribed will be done. Contractors are expected to commit the resources necessary for it to happen.

Stacking of trades—When multiple crews from various trades are assigned to the same area, thereby creating congestion. There are too many workers working in the same area.

Surge pile—An excess pile of materials that a crew can rely upon and use when materials have been or are not being delivered.

Symbiotic crew relationships—When the completion of the work requires multiple crews or subcontractors. In symbiotic work, one crew or worker is often in a standby mode waiting for another crew or worker to advance the progress to the point where the succeeding crew or worker can advance the progress further. Thus, the pace of the crew is established by someone else. An example is when an electrician installs PVC conduit and receptacles in a masonry wall. The electrician's work can proceed no faster than the masonry.

Total cost—A method of quantifying cost overruns by comparing the total cost incurred with the contractor's estimate.

Trade sequencing—The practice of sequencing the work of multiple trade contractors. It is normally discussed in conjunction with interior service and finish work.

Unit rate—Labor productivity measured as the work hours per unit of work.

Velocity (production rate) charts—Graphs that plot the production rate or percentage of output versus time. They are useful in determining time lags.

Work face—The area on the site where the workers install or erect the facility components.

Work hours—Working hours.

Workforce management—The collection of management practices that define how to manage a construction workforce at the crew level to maximize production.

Index

Page numbers followed by *f* or *t* refer to figures or tables.

C

Y

About the Authors

H. Randolph Thomas, Ph.D., P.E. is a retired professor emeritus of civil engineering at the Pennsylvania State University. He received his Ph.D. degree in civil engineering from Vanderbilt University in 1976. For 35 years, Dr. Thomas taught and conducted research in the areas of construction labor productivity and construction contracts. He has written extensively on both topics in various ASCE journals. In 1997, he was awarded a Fulbright scholarship to study construction practices in Eastern Europe. In 2000, he received ASCE's Peurifoy Construction Research Award. He retired as a Fellow member of ASCE and was a registered professional engineer in the state of Tennessee.

Ralph D. Ellis Jr., Ph.D., P.E. is an associate professor of civil engineering at the University of Florida, where he heads the Construction Engineering Management program. He has 15 years of industry experience as a construction project manager and business manager. He received his MBA degree from Nova University in 1987 and his Ph.D. degree in civil engineering from the University of Florida in 1989. Dr. Ellis teaches and performs research in the legal aspects of construction engineering and engineering project management. He is a registered professional engineer in the state of Florida.

About the Authors